AF679553

# Finite Element Method Lecture Notes

## A Student Practice Textbook for Scientists and Engineers

Peter I. Kattan

Petra Books
www.PetraBooks.com

Finite Element Method Lecture Notes
A Student Practice Textbook for Scientists and Engineers

Peter I. Kattan, PhD

*Correspondence about this book may be sent to the author at one of the following two email addresses:*

pkattan@petrabooks.com

info@petrabooks.com

---

Finite Element Method Lecture Notes
ISBN: 979-8-3485-0441-0
All rights reserved. No part of this book may be copied or reproduced without written permission of the author or publisher.
© 2025 Peter I. Kattan

# Preface

These are the handwritten notes of the finite element analysis course that was taught by Dr. Peter I. Kattan at Applied Science University in the period 2000-2002.

These notes are based on the book "A First Course in the Finite Element Method" by Daryl Logan. These note chapters have become very popular with students and it is good to find them in one signle volume.

The author offers two versions of these notes: one free downloadable PDF version and one paid print edition.

The notes cover both one-dimensional elements and two-dimensional elements.

Peter I. Kattan February 2025

Finite Element Method Lecture Notes
A Student Practice Textbook for Scientists and Engineers

# Contents

Chapter 1: Introduction 1

Chapter 2: Trusses 29

Chapter 3: One-Dimensional Elements 42

Chapter 4: Analysis of One-Dimensional Problems 56

Chapter 5: Two-Dimensional Elements 58

# Finite Element Analysis
## Chapter 1
## Introduction

Page 1

**Definition**: The finite element method (FEM) is a numerical procedure that can be used to obtain solutions to a large class of engineering problems involving stress analysis, heat transfer, electromagnetism, and fluid flow.

**Note**: We will use the general-purpose FEM program ANSYS.

**Objective of Chapter**: to introduce the basic concepts of finite element analysis, including:

1. Direct formulation.
2. Minimum potential energy theorem.
3. Weighted residual methods.

---

### 1.1 Engineering Problems:

Page 2

* Engineering problems are mathematical models of physical systems.
* Mathematical models are differential equations with a set of corresponding boundary and initial conditions.

---

Page 2

### 1.2 Numerical Methods:

* There are many practical engineering problems for which we cannot obtain exact solutions.
* In these cases, we resort to numerical approximations.
* The first step of any numerical procedure is discretization.
* This process divides the medium into a number of small subregions

and nodes. The subregions are called elements.

* An approximate continuous function is assumed to represent the solution for each element.
* The complete solution is then generated by connecting or assembling the individual solutions, allowing for continuity at the interelemental boundaries.

---

## Physical Properties Characterizing Various Engineering Systems:

Page 3

### ① Solid Mechanics Examples:

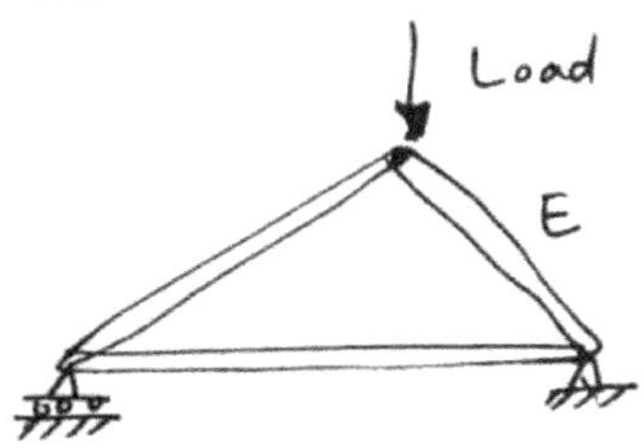


A Truss

E : modulus of elasticity

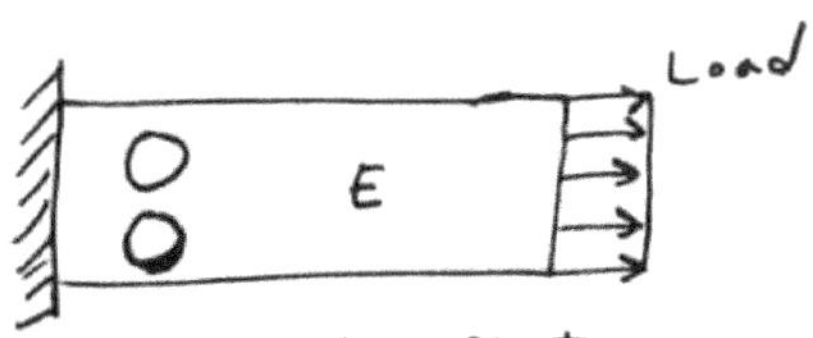


An Elastic Plate

E : modulus of elasticity

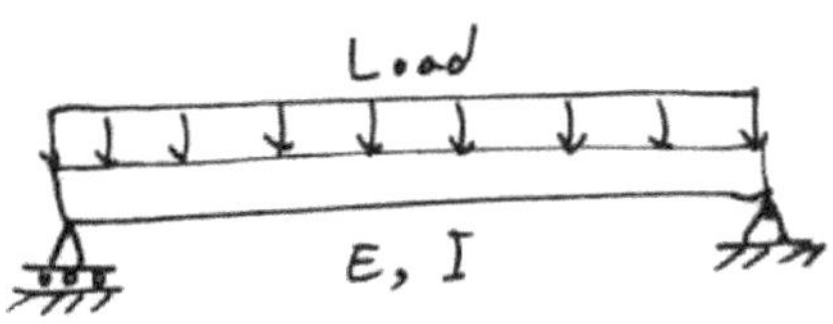


A Beam

E : modulus of elasticity

I : second moment of area

A Shaft

G : modulus of rigidity

J : polar moment of inertia of the area.

② Fluid Flow Examples:

μ : viscosity

e : relative roughness

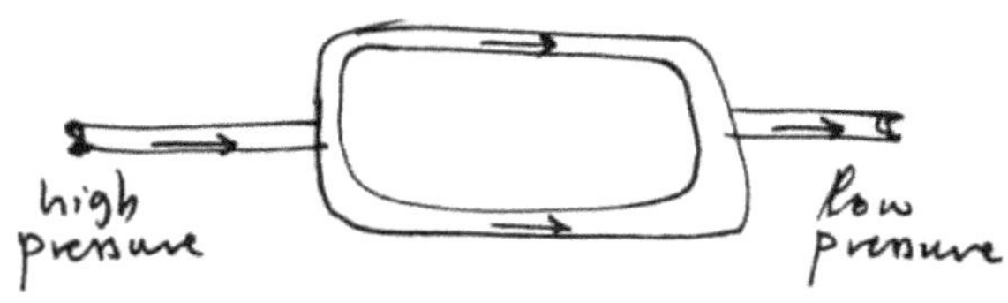


Pipe Networks

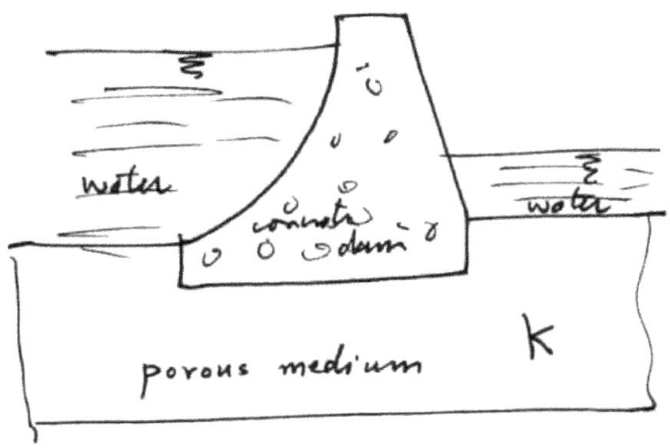


A Concrete Dam

K : soil permeability

---

③ Heat Transfer Examples: (not required).

④ Electrical and Magnetism Examples: (not required).

---

1.3 A Brief History of the Finite Element Method and ANSYS:

Page 5

* Courant (1943) is the first person to develop the finite element method.

* Clough (1960) made the term "finite element" popular.

* Zienkiewicz and Cheung (1967) wrote the first book entirely devoted to the finite element method.

* In 1971, ANSYS was released for the first time.

* ANSYS is a comprehensive general-purpose finite element computer program that contains over 100,000 lines of code.

---

## 1.4 Basic Steps in the Finite Element Method: Page 6

### Preprocessing Phase:

1. Create and discretize the solution domain into finite elements; i.e., subdivide the problem into nodes and elements.

2. Assume a shape function to represent the physical behavior of an element; i.e., an approximate continuous function is assumed to represent the solution of an element.

3. Develop equations for an element.

4. Assemble the elements to present the entire problem. Construct the global stiffness matrix.

5. Apply boundary conditions, initial conditions, and loading.

### Solution Phase:

6. Solve a set of linear or nonlinear algebraic equations simultaneously to obtain nodal results, such as displacement values at different nodes, or temperature values at different nodes in a heat transfer problem.

### Postprocessing Phase:

7. Obtain other important information. At this point, you may be interested in values of principal stresses, heat fluxes, etc.

---

Check the examples of the capabilities of ANSYS on page 7 (or Cover of Book).

---

## 1.5 Direct Formulation:

Page 6

### Example 1.1 :

Page 8

Consider a bar with a variable cross-section supporting a load P as shown in the figure. The bar is fixed at one end and carries the load P at the other end. Let us designate the width of the bar at the top by $w_1$, at the bottom by $w_2$, its thickness by $t$, and its length by $L$. The bar's modulus of elasticity will be denoted by $E$.
Determine how much the bar will deflect at various points along its length when it is subjected to the load P.
Neglect the weight of the bar in the analysis assuming that the applied load is considerably larger than the weight of the bar.

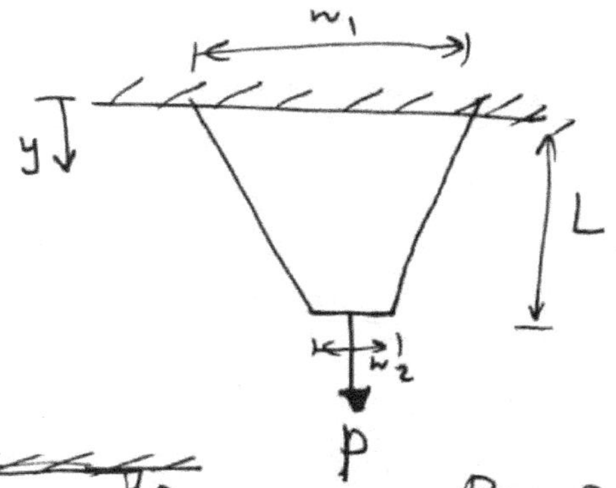


### Solution :

#### Preprocessing Phase :

(1) Discretize the solution domain into finite elements.

Subdivide the bar into
4 elements.
⟹ 5 nodes.

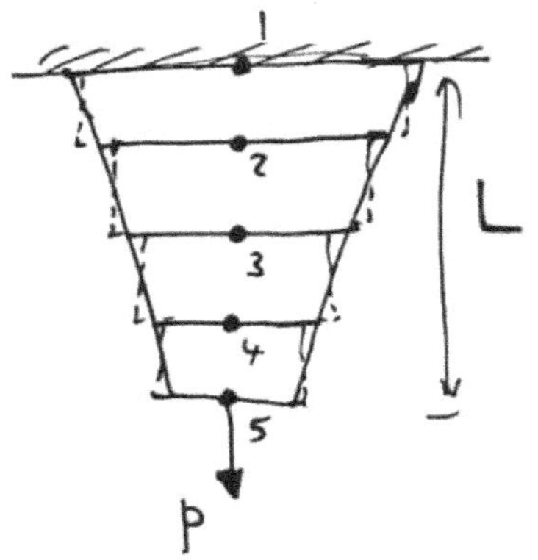


Page 8

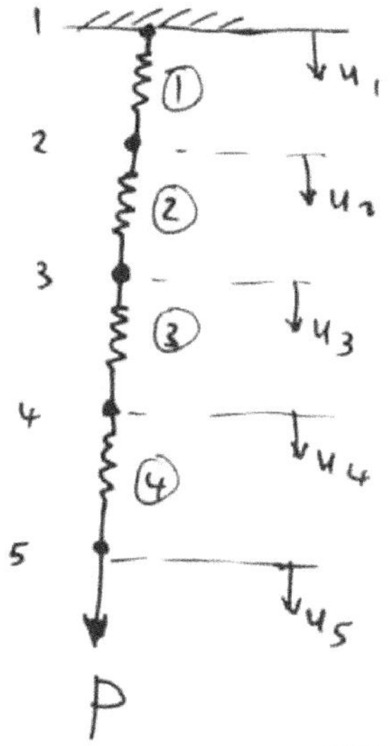

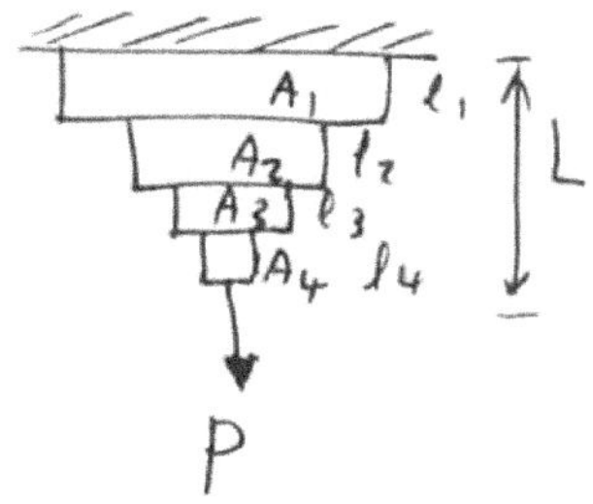

$l_1 + l_2 + l_3 + l_4 = L$

Unknowns:

$u_1, u_2, u_3, u_4, u_5$ : displacements at the nodes.

**Note**: We can increase the accuracy of the results by subdividing the bar into a larger number of elements and nodes.

the cross-sectional area of each element is represented by an average area of the cross-sections at the nodes that make up (define) the element.

---

② Assume a solution that approximates the behavior of an element:

Page 9

For an element with a uniform cross-section:

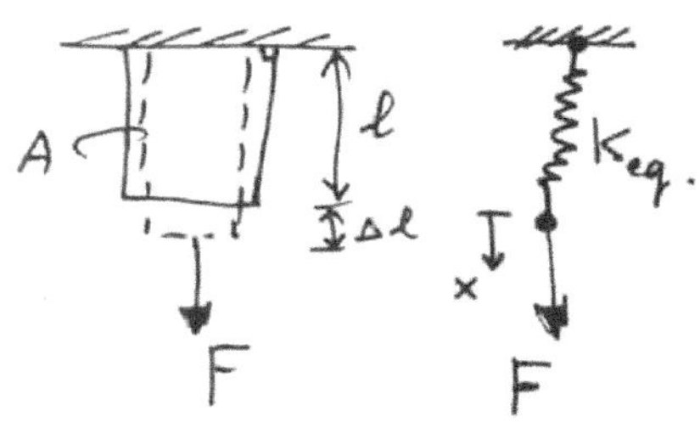

Equivalent System

stress: $\sigma = \frac{F}{A}$

strain: $\varepsilon = \frac{\Delta l}{l}$

Hooke's Law (stress-strain relation): $\sigma = E\varepsilon$

Substitute: $\frac{F}{A} = E\frac{\Delta l}{l}$

$$\Rightarrow \boxed{F = \left(\frac{EA}{l}\right)\Delta l}$$

Equivalent stiffness $\equiv K_{eq.} = K_{equivalent}$

Here, $\boxed{K_{eq.} = \frac{EA}{L}} \Rightarrow \boxed{F = K_{eq.} \Delta l}$

---

Force in each element is $f = K_{eq.}(u_{i+1} - u_i)$

but $K_{eq.} = \frac{EA_{avg.}}{l}$ , where $A_{avg.} = \frac{A_{i+1} + A_i}{2}$

$$\therefore f = \frac{E(A_{i+1} + A_i)}{2l}(u_{i+1} - u_i)$$

$$\Rightarrow \boxed{k_{eq.} = \frac{E(A_{i+1} + A_i)}{2l}} \text{ for this example.}$$

where
- $A_i$ : cross-sectional area at node $i$.
- $A_{i+1}$ : cross-sectional area at node $i+1$.
- $l$ : length of the element.

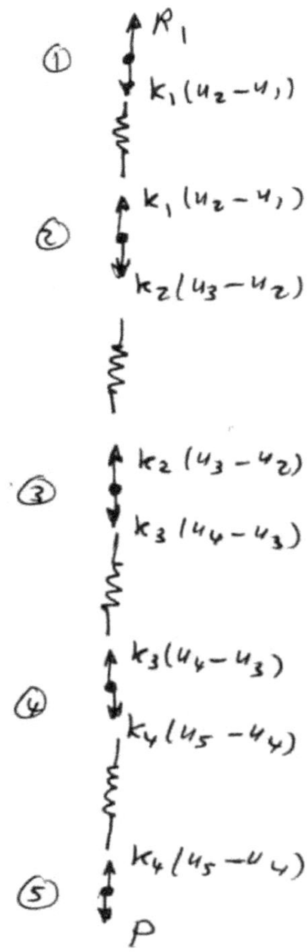


Write equilibrium equations at the <u>five nodes</u> (<u>five equations</u>):

<u>node 1</u>: $R_1 - k_1(u_2 - u_1) = 0$ ——— ①

<u>node 2</u>: $k_1(u_2 - u_1) - k_2(u_3 - u_2) = 0$ ——— ②

<u>node 3</u>: $k_2(u_3 - u_2) - k_3(u_4 - u_3) = 0$ ——— ③

<u>node 4</u>: $k_3(u_4 - u_3) - k_4(u_5 - u_4) = 0$ ——— ④

<u>node 5</u>: $k_4(u_5 - u_4) - P = 0$ ——— ⑤

$$k_1 u_1 - k_1 u_2 = -R_1 \quad \text{——— (1)}$$

$$-k_1 u_1 + (k_1 + k_2) u_2 - k_2 u_3 = 0 \quad \text{——— (2)}$$

$$-k_2 u_2 + (k_2 + k_3) u_3 - k_3 u_4 = 0 \quad \text{——— (3)}$$

$$-k_3 u_3 + (k_3 + k_4) u_4 - k_4 u_5 = 0 \quad \text{——— (4)}$$

$$-k_4 u_4 + k_4 u_5 = P \quad \text{——— (5)}$$

---

Rewrite the equations in matrix form: displacement matrix, load matrix.

$$\begin{bmatrix} k_1 & -k_1 & 0 & 0 & 0 \\ -k_1 & k_1+k_2 & -k_2 & 0 & 0 \\ 0 & -k_2 & k_2+k_3 & -k_3 & 0 \\ 0 & 0 & -k_3 & k_3+k_4 & -k_4 \\ 0 & 0 & 0 & -k_4 & k_4 \end{bmatrix}_{5\times5} \begin{Bmatrix} u_1 \\ u_2 \\ u_3 \\ u_4 \\ u_5 \end{Bmatrix}_{5\times1} = \begin{Bmatrix} -R_1 \\ 0 \\ 0 \\ 0 \\ P \end{Bmatrix}_{5\times1}$$

"Stiffness Matrix"

Banded Symmetric Positive-Definite Matrix.

$$\begin{bmatrix} k_1 & -k_1 & 0 & 0 & 0 \\ -k_1 & k_1+k_2 & -k_2 & 0 & 0 \\ 0 & -k_2 & k_2+k_3 & -k_3 & 0 \\ 0 & 0 & -k_3 & k_3+k_4 & -k_4 \\ 0 & 0 & 0 & -k_4 & k_4 \end{bmatrix}_{5\times5} \begin{Bmatrix} u_1 \\ u_2 \\ u_3 \\ u_4 \\ u_5 \end{Bmatrix}_{5\times1} - \begin{Bmatrix} 0 \\ 0 \\ 0 \\ 0 \\ P \end{Bmatrix}_{5\times1} = \begin{Bmatrix} -R_1 \\ 0 \\ 0 \\ 0 \\ 0 \end{Bmatrix}_{5\times1}$$

$$\boxed{[K]\{u\} - \{F\} = \{R\}}$$

$[K]$: stiffness matrix; $\{u\}$: displacement matrix; $\{F\}$: load matrix; $\{R\}$: reaction matrix

Boundary Conditions : $u_1 = 0$ (at fixed support).

$$
\left[\begin{array}{c|cccc}
k_1 & -k_1 & 0 & 0 & 0 \\
\hline
-k_1 & k_1+k_2 & -k_2 & 0 & 0 \\
0 & -k_2 & k_2+k_3 & -k_3 & 0 \\
0 & 0 & -k_3 & k_3+k_4 & -k_4 \\
0 & 0 & 0 & -k_4 & k_4
\end{array}\right]_{5\times5}
\left\{\begin{array}{c}
0 \\
\hline
u_2 \\ u_3 \\ u_4 \\ u_5
\end{array}\right\}_{5\times1}
=
\left\{\begin{array}{c}
-R_1 \\
\hline
0 \\ 0 \\ 0 \\ P
\end{array}\right\}_{5\times1}
$$

Partition the system as shown above to get :

$$
\underbrace{[k_1][0]_{1\times1}}_{0} + [-k_1 \quad 0 \quad 0 \quad 0]_{1\times5} \begin{Bmatrix} u_2 \\ u_3 \\ u_4 \\ u_5 \end{Bmatrix}_{5\times1} = -R_1 \,{}_{1\times1} \qquad \text{—— (1)}
$$

$$
\underbrace{\begin{Bmatrix} -k_1 \\ 0 \\ 0 \\ 0 \end{Bmatrix}_{5\times1} [0]_{1\times1}}_{0} + \begin{bmatrix} k_1+k_2 & -k_2 & 0 & 0 \\ -k_2 & k_2+k_3 & -k_3 & 0 \\ 0 & -k_3 & k_3+k_4 & -k_4 \\ 0 & 0 & -k_4 & k_4 \end{bmatrix}_{5\times5} \begin{Bmatrix} u_2 \\ u_3 \\ u_4 \\ u_5 \end{Bmatrix}_{5\times1} = \begin{Bmatrix} 0 \\ 0 \\ 0 \\ P \end{Bmatrix}_{5\times1} \qquad \text{—— (2)}
$$

$\Rightarrow$

$$
\boxed{-k_1 u_2 = -R_1} \qquad \text{—— (1)}
$$

$$
\begin{bmatrix} k_1+k_2 & -k_2 & 0 & 0 \\ -k_2 & k_2+k_3 & -k_3 & 0 \\ 0 & -k_3 & k_3+k_4 & -k_4 \\ 0 & 0 & -k_4 & k_4 \end{bmatrix}_{4\times4} \begin{Bmatrix} u_2 \\ u_3 \\ u_4 \\ u_5 \end{Bmatrix}_{4\times1} = \begin{Bmatrix} 0 \\ 0 \\ 0 \\ P \end{Bmatrix}_{4\times1} \qquad \text{—— (2)}
$$

③ **Develop equations for an element:**

$$f_i = k_{eq.}(u_{i+1} - u_i)$$

$$f_{i+1} = k_{eq.}(u_{i+1} - u_i)$$

Note that $\boxed{f_i = f_{i+1}}$

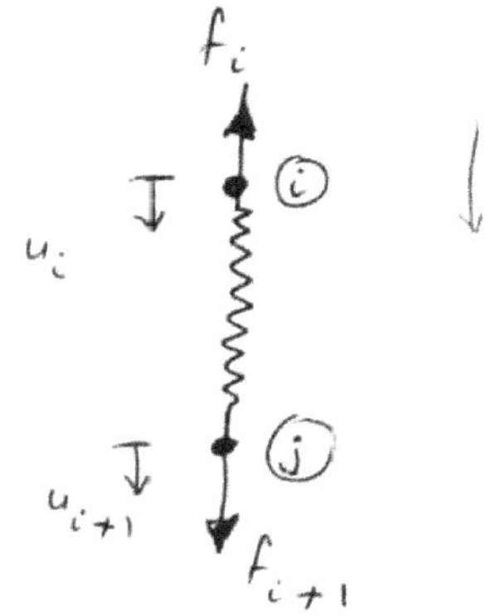


Use the new sign convention:

$$\boxed{\begin{aligned} f_i &= k_{eq.}(u_i - u_{i+1}) \\ f_{i+1} &= k_{eq.}(u_{i+1} - u_i) \end{aligned}}$$

New Sign Convention:

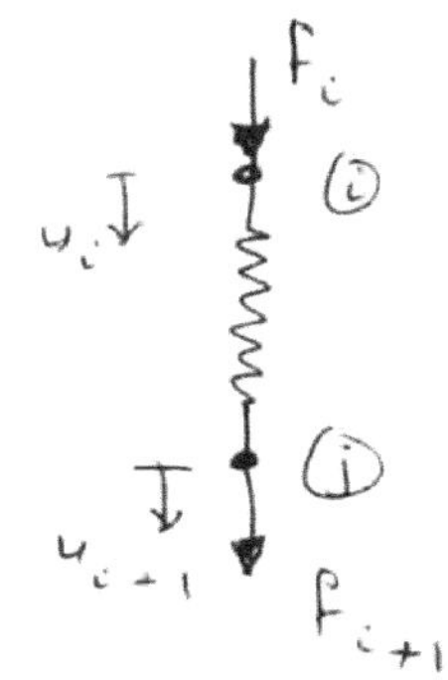


Rewrite in matrix form:

$$\begin{Bmatrix} f_i \\ f_{i+1} \end{Bmatrix} = \begin{bmatrix} k_{eq.} & -k_{eq.} \\ -k_{eq.} & k_{eq.} \end{bmatrix} \begin{Bmatrix} u_i \\ u_{i+1} \end{Bmatrix}$$

"Element stiffness matrix".

④ **Assemble the elements to present the entire problem:**

Direct Formulation

| | 1 | 2 | 3 | 4 | 5 |
|---|---|---|---|---|---|
| 1 | $k_1$ | $-k_1$ | | 0 | 0 |
| 2 | $-k_1$ | $k_1+k_2$ | $-k_2$ | 0 | 0 |
| 3 | 0 | $-k_2$ | $k_2+k_3$ | $-k_3$ | 0 |
| 4 | 0 | 0 | $-k_3$ | $k_3+k_4$ | $-k_4$ |
| 5 | 0 | 0 | 0 | $-k_4$ | $k_4$ |

$5 \times 5$

"Global stiffness matrix"

⑤ Apply boundary conditions:

See equations ① and ② on page 9.

## Solution Phase:

⑥ Solve a system of algebraic equations simultaneously:

Use the following numerical values:

$E = 10.4 \times 10^6 \ \text{lb/in}^2$ (psi) for aluminum.

$w_1 = 2$ in.

$w_2 = 1$ in.

$t = 0.125$ in.

$L = 10$ in.

$P = 1000$ lb $\equiv 1$ kip.

For each element: $l = \frac{L}{4} = \frac{10}{4} = 2.5$ in.

At distance $y$:

$$w = w_1 + \left(\frac{w_2 - w_1}{L}\right) y$$

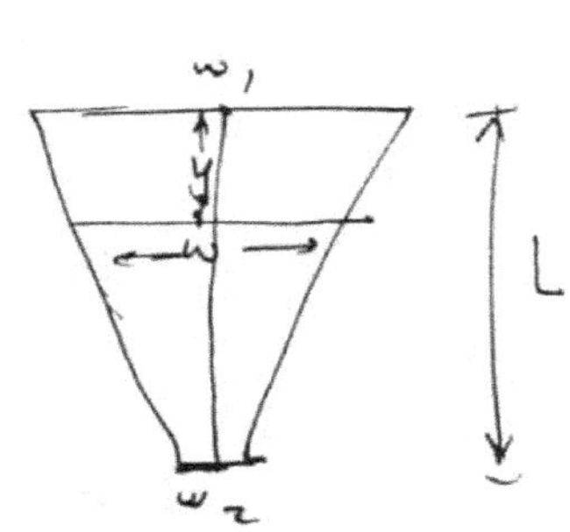


$$A = wt = \left[w_1 + \left(\frac{w_2 - w_1}{L}\right) y\right] t.$$

$$A = \left[2 + \frac{(1-2)}{10} y\right](0.125)$$

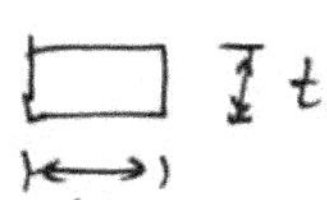


$$\boxed{A = 0.25 - 0.0125\, y} \quad \text{function of } y.$$

$A_1 = 0.25 - 0.0125\,(0) = 0.25 \ \text{in}^2.$

$A_2 = 0.25 - 0.0125\,(2.5) = 0.21875 \ \text{in}^2.$

$A_3 = 0.25 - 0.0125\,(5.0) = 0.1875 \ \text{in}^2.$

$A_4 = 0.25 - 0.0125\,(7.5) = 0.15625 \ \text{in}^2.$

$A_5 = 0.25 - 0.0125\,(10.0) = 0.125 \ \text{in}^2.$

$$K_{eq.} = \frac{(A_{i+1} + A_i)E}{2\ell} \qquad \frac{in^2 \cdot \frac{lb}{in^2}}{in} = \frac{lb}{in}$$

$$k_1 = \frac{(0.25 + 0.21875)(10.4 \times 10^6)}{2(2.5)} = 975 \times 10^3 \ lb/in.$$

$$k_2 = \frac{(0.21875 + 0.1875)(10.4 \times 10^6)}{2(2.5)} = 845 \times 10^3 \ lb/in.$$

$$k_3 = \frac{(0.1875 + 0.15625)(10.4 \times 10^6)}{2(2.5)} = 715 \times 10^3 \ lb/in.$$

$$k_4 = \frac{(0.15625 + 0.125)(10.4 \times 10^6)}{2(2.5)} = 585 \times 10^3 \ lb/in.$$

See table page 14.

The element stiffness matrices are:

$$[k]^{(1)} = \begin{bmatrix} k_1 & -k_1 \\ -k_1 & k_1 \end{bmatrix} = 10^3 \begin{bmatrix} 975 & -975 \\ -975 & 975 \end{bmatrix} \ lb/in.$$

$$[k]^{2} = \begin{bmatrix} k_2 & -k_2 \\ -k_2 & k_2 \end{bmatrix} = 10^3 \begin{bmatrix} 845 & -845 \\ -845 & 845 \end{bmatrix} \ lb/in.$$

$$[k]^{(3)} = \begin{bmatrix} k_3 & -k_3 \\ -k_3 & k_3 \end{bmatrix} = 10^3 \begin{bmatrix} 715 & -715 \\ -715 & 715 \end{bmatrix} \ lb/in.$$

$$[k]^{(4)} = \begin{bmatrix} k_4 & -k_4 \\ -k_4 & k_4 \end{bmatrix} = 10^3 \begin{bmatrix} 585 & -585 \\ -585 & 585 \end{bmatrix} \ lb/in.$$

Assemble the element stiffness matrices to get the global stiffness matrix :

$$[K]^G = 10^3 \begin{bmatrix} 975 & -975 & & & \\ -975 & 975+845 & -845 & & \\ & -845 & 845+715 & -715 & \\ & & -715 & 715+585 & -585 \\ & & & -585 & 585 \end{bmatrix} \text{lb/in.}$$

(rows and columns numbered 1, 2, 3, 4, 5)

Boundary condition: $u_1 = 0$

$$10^3 \begin{bmatrix} 975 & -975 & 0 & 0 & 0 \\ -975 & 1820 & -845 & 0 & 0 \\ 0 & -845 & 1560 & -715 & 0 \\ 0 & 0 & -715 & 1300 & -585 \\ 0 & 0 & 0 & -585 & 585 \end{bmatrix} \begin{Bmatrix} u_1 \\ u_2 \\ u_3 \\ u_4 \\ u_5 \end{Bmatrix} = \begin{Bmatrix} -R_1 \\ 0 \\ 0 \\ 0 \\ P = 1000 \end{Bmatrix}$$

lb

$$\Rightarrow \boxed{-10^3 (+975\, u_2) = -R_1}$$

$$10^3 \begin{bmatrix} 1820 & -845 & 0 & 0 \\ -845 & 1560 & -715 & 0 \\ 0 & -715 & 1300 & -585 \\ 0 & 0 & -585 & 585 \end{bmatrix}_{4\times4} \begin{Bmatrix} u_2 \\ u_3 \\ u_4 \\ u_5 \end{Bmatrix}_{4\times1} = \begin{Bmatrix} 0 \\ 0 \\ 0 \\ 1000 \end{Bmatrix}_{4\times1}$$

Partition again :

$$10^3 \begin{bmatrix} 1820 & -845 & 0 \\ -845 & 1560 & -715 \\ 0 & -715 & 1300 \end{bmatrix} \begin{Bmatrix} u_2 \\ u_3 \\ u_4 \end{Bmatrix} + 10^3 \begin{Bmatrix} 0 \\ 0 \\ -585 \end{Bmatrix} u_5 = \begin{Bmatrix} 0 \\ 0 \\ 0 \end{Bmatrix}$$

$$10^3 \begin{pmatrix} 0 & 0 & -585 \end{pmatrix} \begin{Bmatrix} u_2 \\ u_3 \\ u_4 \end{Bmatrix} + 10^3 (585 u_5) = 1000 \quad \text{——} \; (2)$$

From equation (1) :

$$\begin{bmatrix} 1820 & -845 & 0 \\ -845 & 1560 & -715 \\ 0 & -715 & 1300 \end{bmatrix} \begin{Bmatrix} u_2 \\ u_3 \\ u_4 \end{Bmatrix} = - \begin{Bmatrix} 0 \\ 0 \\ -585 \end{Bmatrix} u_5$$

$$\therefore \begin{Bmatrix} u_2 \\ u_3 \\ u_4 \end{Bmatrix} = \begin{bmatrix} 1820 & -845 & 0 \\ -845 & 1560 & -715 \\ 0 & -715 & 1300 \end{bmatrix}^{-1} \begin{Bmatrix} 0 \\ 0 \\ 585 \end{Bmatrix} u_5$$

Find the inverse matrix by Gaussian elimination :

$$\left(\frac{845}{1820}\right) * \left[\begin{array}{ccc|ccc} 1820 & -845 & 0 & 1 & 0 & 0 \\ -845 & 1560 & -715 & 0 & 1 & 0 \\ 0 & -715 & 1300 & 0 & 0 & 1 \end{array}\right]$$

$$\left(\frac{715}{1168}\right) * \left[\begin{array}{ccc|ccc} 1820 & -845 & 0 & 1 & 0 & 0 \\ 0 & 1168 & -715 & 0.464 & 1 & 0 \\ 0 & -715 & 1300 & 0 & 0 & 1 \end{array}\right]$$

$$\left(\frac{845}{1168}\right) * \left[\begin{array}{ccc|ccc} 1820 & -845 & 0 & 1 & 0 & 0 \\ 0 & 1168 & -715 & 0.464 & 1 & 0 \\ 0 & 0 & 862 & 0.284 & 0.612 & 1 \end{array}\right]$$

$$\left(\frac{517}{862}\right) * \left[\begin{array}{ccc|ccc} 1820 & 0 & -517 & 1.336 & 0.723 & 0 \\ 0 & 1168 & -715 & 0.464 & 1 & 0 \\ 0 & 0 & 862 & 0.284 & 0.612 & 1 \end{array}\right]$$

$$\left(\frac{715}{862}\right) * \left[\begin{array}{ccc|ccc} 1820 & 0 & 0 & 1.506 & 1.090 & 0.600 \\ 0 & 1168 & -715 & 0.464 & 1 & 0 \\ 0 & 0 & 862 & 0.284 & 0.612 & 1 \end{array}\right]$$

$$\begin{array}{c} \frac{1}{1820} * \\ \frac{1}{1168} * \\ \frac{1}{862} * \end{array} \left[\begin{array}{ccc|ccc} 1820 & 0 & 0 & 1.506 & 1.090 & 0.6 \\ 0 & 1168 & 0 & 0.7 & 1.508 & 0.829 \\ 0 & 0 & 862 & 0.284 & 0.612 & 1 \end{array}\right]$$

$$\left[\begin{array}{ccc|ccc} 1 & 0 & 0 & 0.000827 & 0.000599 & 0.000330 \\ 0 & 1 & 0 & 0.000599 & 0.001291 & 0.0007098 \\ 0 & 0 & 1 & 0.000329 & 0.0007098 & 0.001160 \end{array}\right]$$

$$\Rightarrow \text{Inverse matrix} \equiv 10^{-3} \begin{bmatrix} 0.827 & 0.599 & 0.330 \\ 0.599 & 1.291 & 0.710 \\ 0.330 & 0.710 & 1.160 \end{bmatrix}$$

$$\therefore \begin{Bmatrix} u_2 \\ u_3 \\ u_4 \end{Bmatrix} = 10^{-3} \begin{bmatrix} 0.827 & 0.599 & 0.330 \\ 0.599 & 1.291 & 0.710 \\ 0.330 & 0.710 & 1.160 \end{bmatrix} \begin{Bmatrix} 0 \\ 0 \\ 585 \end{Bmatrix} u_5$$

$$= 10^{-3} \begin{Bmatrix} 193.05 \\ 415.35 \\ 678.6 \end{Bmatrix} u_5$$

Substitute in eq. ②:

$$10^{3} \begin{bmatrix} 0 & 0 & -585 \end{bmatrix} \cdot 10^{-3} \begin{Bmatrix} 193.05 \\ 415.35 \\ 678.6 \end{Bmatrix} u_5 + 10^3 (585\, u_5) = 1000$$

$$-585\,(678.6)\, u_5 + 10^3 (585\, u_5) = 1000$$

$$\Longrightarrow \quad 188019\, u_5 = 1000$$

$$\Longrightarrow \quad \boxed{u_5 = 0.0053186 \text{ in}}.$$

0.005317 in. in textbook

$$\therefore \begin{Bmatrix} u_2 \\ u_3 \\ u_4 \end{Bmatrix} = 10^{-3} \begin{Bmatrix} 193.05 \\ 415.35 \\ 678.6 \end{Bmatrix} (0.0053186) = \begin{Bmatrix} 0.001027 \\ 0.002209 \\ 0.0036092 \end{Bmatrix} \text{ in.}$$

$$\Longrightarrow \quad \boxed{\begin{aligned} u_2 &= 0.001027 \text{ in.} \\ u_3 &= 0.002209 \text{ in.} \\ u_4 &= 0.0036092 \text{ in.} \end{aligned}}$$

## Postprocessing Phase:

**(7) Obtain other information:**

Average Normal Stress in each element:

$$\sigma = \frac{f}{A_{avg.}} \qquad \text{(} f \equiv \text{force in element)}$$

$$= \frac{k_{eq.}\,(u_{i+1} - u_i)}{A_{avg.}}, \quad \text{but } k_{eq.} = \frac{E\,(A_i + A_{i+1})}{2\ell} = \frac{E\,A_{avg.}}{\ell}$$

$$= \frac{E\,A_{avg.}}{L} \cdot \frac{(u_{i+1} - u_i)}{A_{avg.}}$$

$$\boxed{\sigma = E\left(\frac{u_{i+1} - u_i}{\ell}\right)}$$

(psi)

$$\sigma^{(1)} = E\left(\frac{u_2 - u_1}{\ell}\right) = 10.4\times10^6 \left(\frac{0.001027 - 0}{2.5}\right) = 4272 \text{ lb/in}^2.$$

$$\sigma^{(2)} = E\left(\frac{u_3 - u_2}{\ell}\right) = 10.4\times10^6 \left(\frac{0.002209 - 0.001027}{2.5}\right) = 4917 \text{ lb/in}^2.$$

$$\sigma^{(3)} = E\left(\frac{u_4 - u_3}{\ell}\right) = 10.4\times10^6 \left(\frac{0.0036092 - 0.002209}{2.5}\right) = 5825 \text{ lb/in}^2.$$

$$\sigma^{(4)} = E\left(\frac{u_5 - u_4}{\ell}\right) = 10.4\times10^6 \left(\frac{0.0053186 - 0.0036092}{2.5}\right) = 7111 \text{ lb/in}^2.$$

At any section, $f = P = 1000$ lb.

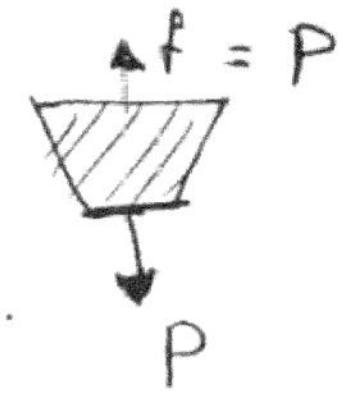


$$A^{(1)}_{avg.} = \frac{A_1 + A_2}{2} = \frac{0.25 + 0.21875}{2} = 0.234375 \text{ in}^2.$$

$$\sigma^{(1)} = \frac{f}{A^{(1)}_{avg.}} = \frac{1000}{0.234375} = 4267 \text{ lb/in}^2 \ (4272).$$

$$A^{(2)}_{avg.} = \frac{0.21875 + 0.1875}{2} = 0.203125 \text{ in}^2.$$

$$\sigma^{(2)} = \frac{f}{A^{(2)}_{avg.}} = \frac{1000}{0.203125} = 4923 \text{ lb/in}^2 \ (4917).$$

$$A^{(3)}_{avg.} = \frac{0.1875 + 0.15625}{2} = 0.171875 \text{ in}^2.$$

$$\sigma^{(3)} = \frac{f}{A^{(3)}_{avg.}} = \frac{1000}{0.171875} = 5818 \text{ lb/in}^2 \ (5829).$$

$$A^{(4)}_{avg.} = \frac{0.15625 + 0.125}{2} = 0.140625 \text{ in}^2.$$

$$\sigma^{(4)} = \frac{f}{A^{(4)}_{avg.}} = \frac{1000}{0.140625} = 7111 \text{ lb/in}^2 \ (7111).$$

**Reactions:**

$$-10^3 (975\, u_2) = -R_1$$

$$\Rightarrow R_1 = 10^3 (975)(0.001027) = 1001 \text{ lb} \approx 1000 \text{ lb}.$$

From statics: $R_1 = P = 1000$ lb.

$$\{R\} = [K]\{u\} - \{F\}$$

$$\begin{Bmatrix} R_1 \\ R_2 \\ R_3 \\ R_4 \\ R_5 \end{Bmatrix} = 10^3 \begin{bmatrix} 975 & -975 & 0 & 0 & 0 \\ -975 & 1820 & -845 & 0 & 0 \\ 0 & -845 & 1560 & -715 & 0 \\ 0 & 0 & -715 & 1300 & -585 \\ 0 & 0 & 0 & -585 & 585 \end{bmatrix} \begin{Bmatrix} 0 \\ 0.001027 \\ 0.002209 \\ 0.0036092 \\ 0.0053186 \end{Bmatrix} - \begin{Bmatrix} 0 \\ 0 \\ 0 \\ 0 \\ 1000 \end{Bmatrix}$$

$$\therefore \begin{Bmatrix} R_1 \\ R_2 \\ R_3 \\ R_4 \\ R_5 \end{Bmatrix} = \begin{Bmatrix} -1001 \cong 1000 \\ 2.5 \cong 0 \\ -2.3 \cong 0 \\ 1.14 \cong 0 \\ -0.001 \cong 0 \end{Bmatrix} = \begin{Bmatrix} -1000 \\ 0 \\ 0 \\ 0 \\ 0 \end{Bmatrix} \text{ lb.}$$

---

**Example 1.2**: (omitted). page 18.

---

**Example 1.3**: A Torsional Problem: Direct Formulation: Page 27

Consider the torsion of circular shafts, shown in the figure. The angle of twist $\theta$ for a shaft with a constant cross-sectional area with a polar moment of inertia $J$ and length $l$, made of homogeneous material with a shear modulus of elasticity $G$, subject to a torque $T$ is given by:

$$\theta = \frac{Tl}{GJ}$$

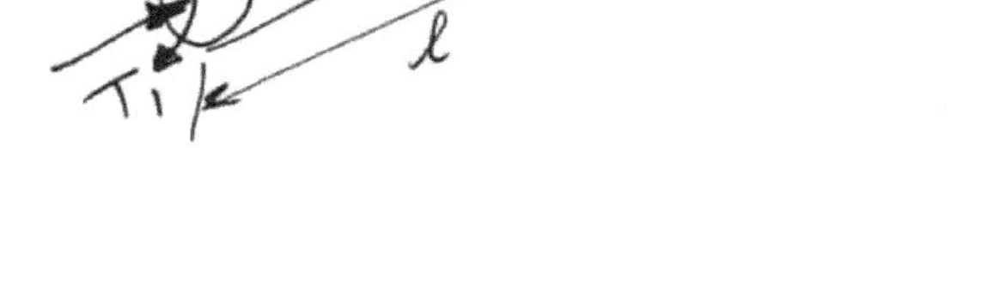

For an element, $\theta = \frac{Tl}{GJ}$

> $\theta$ : angle of twist
> T : torsion.

$$\begin{Bmatrix} T_1 \\ T_2 \end{Bmatrix} = \begin{bmatrix} k & -k \\ -k & k \end{bmatrix} \begin{Bmatrix} \theta_1 \\ \theta_2 \end{Bmatrix}$$

when $k_{eq.} = \frac{GJ}{l}$ (instead of $\frac{EA}{l}$).

Example: A shaft made of two parts.

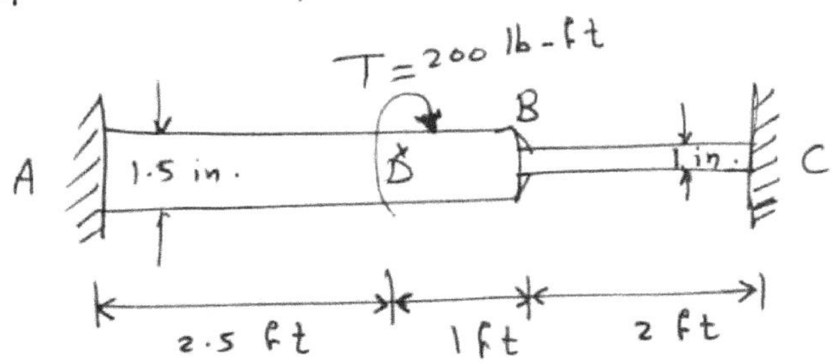


Fixed at both ends.

Part AB: $G_{AB} = 3.9 \times 10^6$ lb/in².
diameter = 1.5 in.

Part BC: $G_{BC} = 4.0 \times 10^6$ lb/in².
diameter = 1 in.

Using three elements, determine the angle of twist at D and B, and the torsional reactions at the boundaries.

Solution: 3 elements
4 nodes.

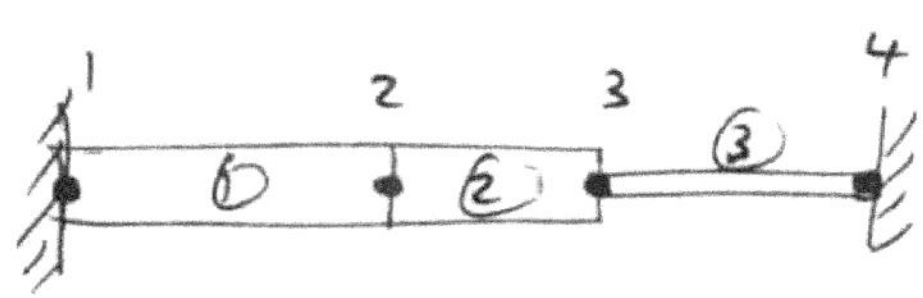


For each element:

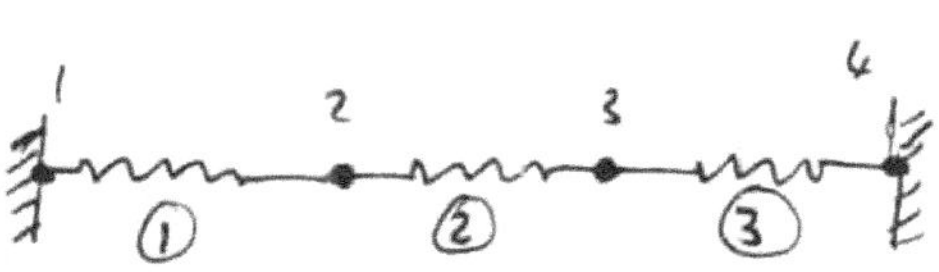


$\boxed{J = \frac{1}{2}\pi r^4}$. Polar moment of inertia.

For each element:

$$J_1 = \frac{1}{2}\pi\left(\frac{1.5}{2}\right)^4 = 0.497 \quad in^4.$$

$$J_2 = \frac{1}{2}\pi\left(\frac{1.5}{2}\right)^4 = 0.497 \quad in^4.$$

$$J_3 = \frac{1}{2}\pi\left(\frac{1}{2}\right)^4 = 0.0982 \quad in^4.$$

$$k_1 = \frac{G_1 J_1}{l_1} = \frac{(3.9\times10^6)(0.497)}{(2.5\times12)} = 64610 \quad lb\cdot in.$$

$$k_2 = \frac{G_2 J_2}{l_2} = \frac{(3.9\times10^6)(0.497)}{(1\times12)} = 161525 \quad lb\cdot in.$$

$$k_3 = \frac{G_3 J_3}{l_3} = \frac{(4.0\times10^6)(0.0982)}{(2\times12)} = 16367 \quad lb\cdot in.$$

Element stiffness matrices:

$$[k]^{(1)} = \begin{bmatrix} k_1 & -k_1 \\ -k_1 & k_1 \end{bmatrix} = \begin{bmatrix} 64610 & -64610 \\ -64610 & 64610 \end{bmatrix} \quad lb-in.$$

$$[k]^{(2)} = \begin{bmatrix} k_2 & -k_2 \\ -k_2 & k_2 \end{bmatrix} = \begin{bmatrix} 161525 & -161525 \\ -161525 & 161525 \end{bmatrix} \quad lb-in.$$

$$[k]^{(3)} = \begin{bmatrix} k_3 & -k_3 \\ -k_3 & k_3 \end{bmatrix} = \begin{bmatrix} 16367 & -16367 \\ -16367 & 16367 \end{bmatrix} \quad lb-in.$$

Assembly Process: $[k]$ is a $4\times4$ matrix.

$$[K] = \begin{array}{c} \begin{array}{cccc} 1 & 2 & 3 & 4 \end{array} \\ \left[\begin{array}{c|c|c|c} 64610 & -64610 & & \\ \hline -64610 & \begin{array}{c}64610\\+161525\end{array} & -161525 & \\ \hline & -161525 & \begin{array}{c}161525\\+16367\end{array} & -16367 \\ \hline & & -16367 & 16367 \end{array}\right] \end{array} \begin{array}{c} \\ 1 \\ 2 \\ 3 \\ 4 \end{array} \quad lb-in.$$

Element load vectors:

200 lb-ft = 200×12 lb-in.
counterclockwise is negative

$$f^{(1)} = \begin{Bmatrix} T_A \\ -200\times12 \end{Bmatrix}$$

$$f^{(2)} = \begin{Bmatrix} 0 \\ 0 \end{Bmatrix} \Longrightarrow \text{global load vector } \{f\} = \begin{Bmatrix} T_A \\ -200\times12 \\ 0 \\ T_c \end{Bmatrix}_{4\times1}$$

$$f^{(3)} = \begin{Bmatrix} 0 \\ T_c \end{Bmatrix}$$

$$\begin{bmatrix} 64610 & -64610 & 0 & 0 \\ -64610 & 226135 & -161525 & 0 \\ 0 & -161525 & 177892 & -16367 \\ 0 & 0 & -16367 & 16367 \end{bmatrix} \begin{Bmatrix} \theta_1 \\ \theta_2 \\ \theta_3 \\ \theta_4 \end{Bmatrix} = \begin{Bmatrix} T_A \\ -200\times12 \\ 0 \\ T_c \end{Bmatrix}$$

Boundary conditions:

$$\boxed{\begin{aligned} \theta_1 &= 0 \\ \theta_4 &= 0 \end{aligned}}$$

$$\left[\begin{array}{c|cc|c} 64610 & -64610 & 0 & 0 \\ \hline -64610 & 226135 & -161525 & 0 \\ 0 & -161525 & 177892 & -16367 \\ \hline 0 & 0 & -16367 & 16367 \end{array}\right] \begin{Bmatrix} 0 \\ \hline \theta_2 \\ \theta_3 \\ \hline 0 \end{Bmatrix} = \begin{Bmatrix} T_A \\ \hline -200\times12 \\ 0 \\ \hline T_c \end{Bmatrix}$$

Partitioning yields:

$$64610[0] + \begin{bmatrix}-64610 & 0\end{bmatrix}\begin{Bmatrix}\theta_2\\ \theta_3\end{Bmatrix} + (0)(0) = T_A$$

$$\Rightarrow \boxed{-64610\,\theta_2 = T_A} \quad \text{—— (1)}$$

$$\begin{Bmatrix}-64610\\ 0\end{Bmatrix}(0) + \begin{bmatrix}226135 & -161525\\ -161525 & 177892\end{bmatrix}\begin{Bmatrix}\theta_2\\ \theta_3\end{Bmatrix} + \begin{Bmatrix}0\\ -16367\end{Bmatrix}(0) = \begin{Bmatrix}-200\times 12\\ 0\end{Bmatrix}$$

$$\Rightarrow \begin{bmatrix}226135 & -161525\\ -161525 & 177892\end{bmatrix}\begin{Bmatrix}\theta_2\\ \theta_3\end{Bmatrix} = \begin{Bmatrix}-200\times 12\\ 0\end{Bmatrix} \quad \text{—— (2)}$$

$$(0)(0) + \begin{bmatrix}0 & -16367\end{bmatrix}\begin{Bmatrix}\theta_2\\ \theta_3\end{Bmatrix} + 16367(0) = T_C$$

$$\boxed{-16367\,\theta_3 = T_C} \quad \text{—— (3)}$$

Use Gaussian elimination.

$$\left(\frac{161525}{226135}\right)* \left[\begin{array}{cc|c}226135 & -161525 & -2400\\ -161525 & 177892 & 0\end{array}\right]$$

$$\left[\begin{array}{cc|c}226135 & -161525 & -2400\\ 0 & 62517 & -1714\end{array}\right]$$

$$\Rightarrow 62517\,\theta_3 = -1714 \Rightarrow \boxed{\theta_3 = -0.02742 \text{ rad}}$$

$$226135\,\theta_2 - 161525(-0.02742) = -2400$$

$$\Rightarrow \boxed{\theta_2 = -0.03020 \text{ rad}}$$

$$\therefore \begin{Bmatrix} \theta_1 \\ \theta_2 \\ \theta_3 \\ \theta_4 \end{Bmatrix} = \begin{Bmatrix} 0 \\ -0.03020 \\ -0.02742 \\ 0 \end{Bmatrix} \text{rad.}$$

$$\Rightarrow T_A = -64610\,\theta_2 = -64610(-0.03020) = 1951 \text{ lb-in.}$$

$$T_c = -16367\,\theta_3 = -16367(-0.02742) = 449 \text{ lb-in.}$$

$$\Rightarrow \{R\} = [K]\{\theta\} - \{T\}$$

$$\therefore \begin{Bmatrix} R_A \\ R_B \\ R_C \\ R_D \end{Bmatrix} = \begin{bmatrix} 64610 & -64610 & 0 & 0 \\ -64610 & 228135 & -161525 & 0 \\ 0 & -161525 & 177892 & -16367 \\ 0 & 0 & -16367 & 16367 \end{bmatrix} \begin{Bmatrix} 0 \\ -0.03020 \\ -0.02742 \\ 0 \end{Bmatrix} - \begin{Bmatrix} 0 \\ -2400 \\ 0 \\ 0 \end{Bmatrix}$$

$$= \begin{Bmatrix} 1951 \\ 0 \\ 0 \\ 449 \end{Bmatrix} \text{lb-in.}$$ (both reactions are clockwise).

<u>Check</u>: $R_A + R_D = 1951 + 449 = \underline{2400}$ lb-in.

---

Page 30

<u>Example 1.4</u>:

A steel plate is subjected to an axial load, as shown in the figure. Approximate the deflections and average stresses along the plate. The plate is $\frac{1}{16}$ in thick and has a modulus of elasticity $E = 29 \times 10^6$ lb/in$^2$.

$t = \frac{1}{16}$ in.

Use 4 elements and 4 nodes.

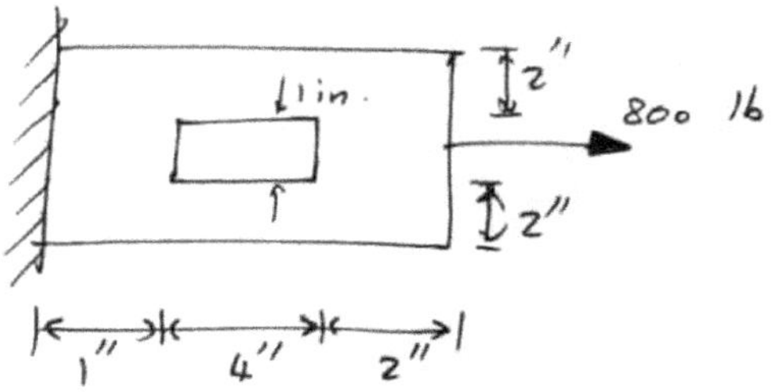


We have axial deformation:

$\therefore\ K_{eq.} = \dfrac{EA}{\ell}$

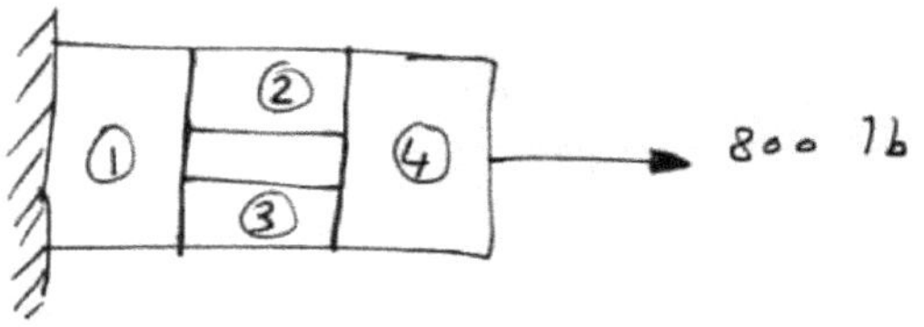


For each element:

$$k_1 = \frac{EA_1}{\ell_1} = \frac{(29\times10^6)(5)(\frac{1}{16})}{1}$$

$$= 9{,}062{,}500 \text{ lb/in.}$$

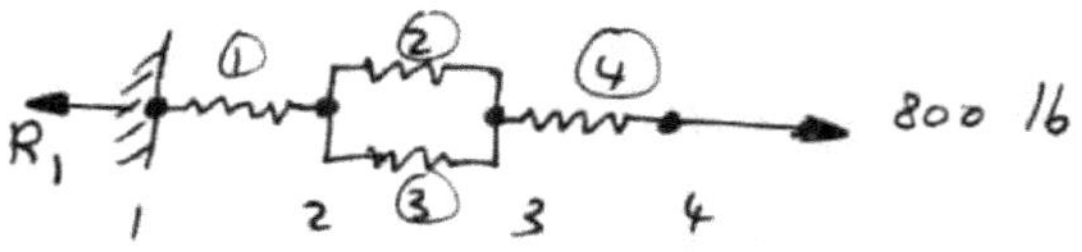


$\longrightarrow x$ is positive to the right.

$$k_2 = \frac{EA_2}{\ell_2} = \frac{(29\times10^6)(2)(\frac{1}{16})}{4}$$

$$= 906{,}250 \text{ lb/in.}$$

$$k_3 = k_2 = 906{,}250 \text{ lb/in.}$$

$$k_4 = \frac{EA_4}{\ell_4} = \frac{(29\times10^6)(5)(\frac{1}{16})}{2} = 4{,}531{,}250 \text{ lb/in.}$$

$$\therefore [k]^{(1)} = \begin{bmatrix} k_1 & -k_1 \\ -k_1 & k_1 \end{bmatrix} = \begin{array}{c} \begin{array}{cc} 1 & 2 \end{array} \\ \begin{bmatrix} 9{,}062{,}500 & -9{,}062{,}500 \\ -9{,}062{,}500 & 9{,}062{,}500 \end{bmatrix} \end{array} \begin{array}{c} 1 \\ 2 \end{array} \text{ lb/in.}$$

$$[k]^{(2)} = \begin{array}{c} \begin{array}{cc} 2 & 3 \end{array} \\ \begin{bmatrix} 906{,}250 & -906{,}250 \\ -906{,}250 & 906{,}250 \end{bmatrix} \end{array} \begin{array}{c} 2 \\ 3 \end{array} \text{ lb/in.}$$

$$[k]^{(3)} = \begin{array}{c} \begin{array}{cc} 2 & 3 \end{array} \\ \begin{bmatrix} 906{,}250 & -906{,}250 \\ -906{,}250 & 906{,}250 \end{bmatrix} \end{array} \begin{array}{c} 2 \\ 3 \end{array} \text{ lb/in.}$$

$$[k]^{(4)} = \begin{array}{c} \begin{array}{cc} 3 & 4 \end{array} \\ \begin{bmatrix} 4{,}531{,}250 & -4{,}531{,}250 \\ -4{,}531{,}250 & 4{,}531{,}250 \end{bmatrix} \end{array} \begin{array}{c} 3 \\ 4 \end{array} \text{ lb/in.}$$

Assembly Process:

$$
[K]_{4\times 4} = \begin{array}{c} \\ \\ \\ \\ \\ \end{array}
\begin{array}{cccc|c}
1 & 2 & 3 & 4 & \\
\hline
9{,}062{,}500 & -9{,}062{,}500 & & & 1 \\
-9{,}062{,}500 & \begin{array}{c}9{,}062{,}500\\+906{,}250\\+906{,}250\end{array} & \begin{array}{c}-906{,}250\\-906{,}250\end{array} & & 2 \\
 & \begin{array}{c}-906{,}250\\-906{,}250\end{array} & \begin{array}{c}906{,}250\\+906{,}250\\+4{,}531{,}250\end{array} & -4{,}531{,}250 & 3 \\
 & & -4{,}531{,}250 & 4{,}531{,}250 & 4
\end{array}
\quad lb/in.
$$

$$[K]\{u\} = \{F\}$$

$$
\begin{bmatrix}
9{,}062{,}500 & -9{,}062{,}500 & 0 & 0 \\
-9{,}062{,}500 & 10{,}875{,}000 & -1{,}812{,}500 & 0 \\
0 & -1{,}812{,}500 & 6{,}343{,}750 & -4{,}531{,}250 \\
0 & 0 & -4{,}531{,}250 & 4{,}531{,}250
\end{bmatrix}
\begin{Bmatrix} \overset{0}{u_1} \\ u_2 \\ u_3 \\ u_4 \end{Bmatrix}
=
\begin{Bmatrix} -R_1 \\ 0 \\ 0 \\ 800 \end{Bmatrix}
$$

Apply boundary conditions: $\boxed{u_1 = 0}$

Partitioning yields:

$$\boxed{-9{,}062{,}500\, u_2 = -R_1} \quad \text{—— (1)}$$

$$
\begin{bmatrix}
10{,}875{,}000 & -1{,}812{,}500 & 0 \\
-1{,}812{,}500 & 6{,}343{,}750 & -4{,}531{,}250 \\
0 & -4{,}531{,}250 & 4{,}531{,}250
\end{bmatrix}
\begin{Bmatrix} u_2 \\ u_3 \\ u_4 \end{Bmatrix}
=
\begin{Bmatrix} 0 \\ 0 \\ 800 \end{Bmatrix}
\quad \text{—— (2)}
$$

Solve by Gaussian elimination:

$$
\left(\frac{1{,}812{,}500}{10{,}875{,}000}\right) *
\left[\begin{array}{ccc|c}
10{,}875{,}000 & -1{,}812{,}500 & 0 & 0 \\
-1{,}812{,}500 & 6{,}343{,}750 & -4{,}531{,}250 & 0 \\
0 & -4{,}531{,}250 & 4{,}531{,}250 & 800
\end{array}\right]
$$

$$\frac{4,531,250}{6,041,667}\ast\left[\begin{array}{ccc|c} 10,875,000 & -1,812,500 & 0 & 0 \\ 0 & 6,041,667 & -4,531,250 & 0 \\ 0 & 0 & 1,132,813 & 800 \end{array}\right]$$

$$\Rightarrow \quad 1,132,813\, u_4 = 800 \Rightarrow u_4 = 7.062\times10^{-4} \text{ in.}$$

$$\Rightarrow \quad 6,041,667\, u_3 - 4,531,250\,(7.062\times10^{-4}) = 0$$

$$\Rightarrow \quad u_3 = 5.297\times10^{-4} \text{ in.}$$

$$10,875,000\, u_2 - 1,812,500\,(5.297\times10^{-4}) = 0$$

$$\Rightarrow \quad u_2 = 8.828\times10^{-5} \text{ in.}$$

$$\Rightarrow \begin{Bmatrix} u_1 \\ u_2 \\ u_3 \\ u_4 \end{Bmatrix} = \begin{Bmatrix} 0 \\ 8.828\times10^{-5} \\ 5.297\times10^{-4} \\ 7.062\times10^{-4} \end{Bmatrix} \text{ in.}$$

**Reaction:** $R_1 = 9,062,500\, u_2 = 9,062,500\,(8.828\times10^{-5}) = 800 \text{ lb.}$

**Element Stresses:** $\sigma = \dfrac{E(u_{i+1} - u_i)}{\ell}$

$$\sigma^{(1)} = \frac{E(u_2 - u_1)}{\ell} = \frac{(29\times10^6)(8.828\times10^{-5} - 0)}{1} = 2560 \text{ lb/in}^2.$$

$$\sigma^{(2)} = \frac{E(u_3 - u_2)}{\ell} = \frac{(29\times10^6)(5.297\times10^{-4} - 8.828\times10^{-5})}{4} = 3200 \text{ lb/in}^2.$$

$$\sigma^{(3)} = \sigma^{(2)} = 3200 \text{ lb/in}^2.$$

$$\sigma^{(4)} = \frac{E(u_4 - u_3)}{\ell} = \frac{(29\times10^6)(7.062\times10^{-4} - 5.297\times10^{-4})}{2} = 2559 \text{ lb/in}^2 \approx 2560 \text{ lb/in}^2.$$

1.6 Minimum Total Potential Energy Formulation: (not required). Page 33.

1.7 Weighted Residual Formulations: (not required). page 38

1.8 Verification of Results: (page 43)

Various sources of error in using FEM software like ANSYS:

① Wrong input data as physical properties and dimensions.

② Selecting inappropriate types of elements.

③ Poor element shape and size after meshing.

④ Applying wrong boundary conditions and loads.

---

Homework 1    pages 45 – 53.

problem
- 6
- 22
- 24

---

# Chapter 2
# Trusses

page 54

## 2.1 Definition of a Truss:

page 54

* A truss is an engineering structure consisting of straight members connected at their ends by means of bolts, rivets, pins, or welding.
* A plane truss is a truss whose members lie in a single plane. The forces acting on such a truss must also lie in this plane.
* Members of a truss are generally considered as two-force members.
* All loads must be applied at the joints.

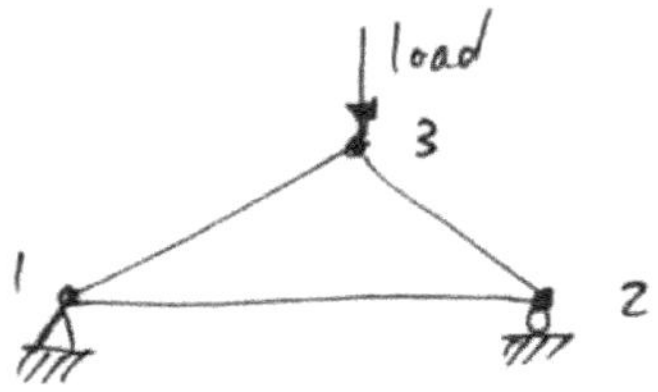


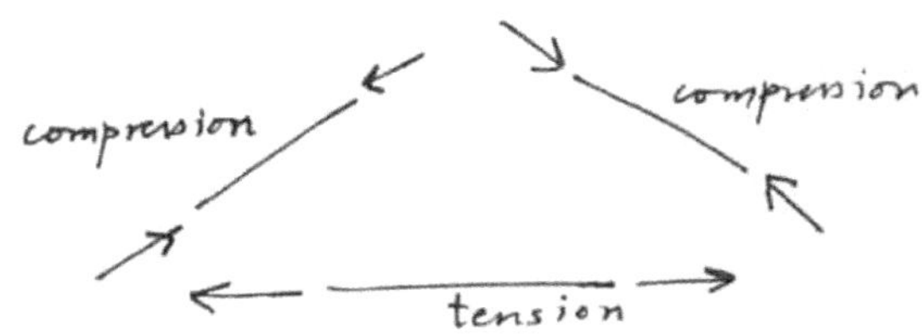


## 2.2 Finite Element Formulation:

page 55

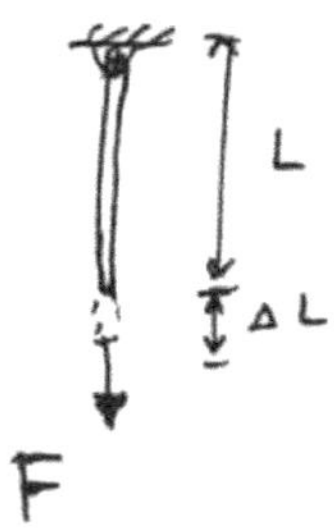


$$\sigma = \frac{F}{A} \quad , \quad \varepsilon = \frac{\Delta L}{L}$$

$F = kx$

$$\sigma = E\varepsilon$$

$$\frac{F}{A} = E \cdot \frac{\Delta L}{L} \Rightarrow F = \left(\frac{EA}{L}\right) \Delta L$$

($\frac{EA}{L}$ = $k_{eq.}$)

$$\Rightarrow \boxed{k_{eq.} = \frac{EA}{L}}$$

(similar to a spring).

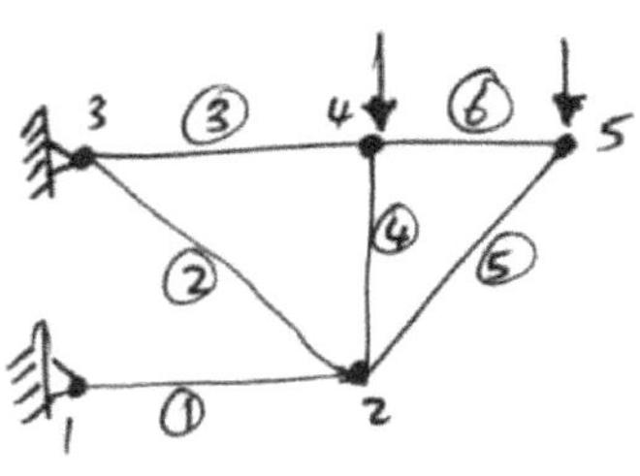


Consider element ⑤:

1- Global coordinate system, $(X, Y)$.

2- Local coordinate system, $(x, y)$.

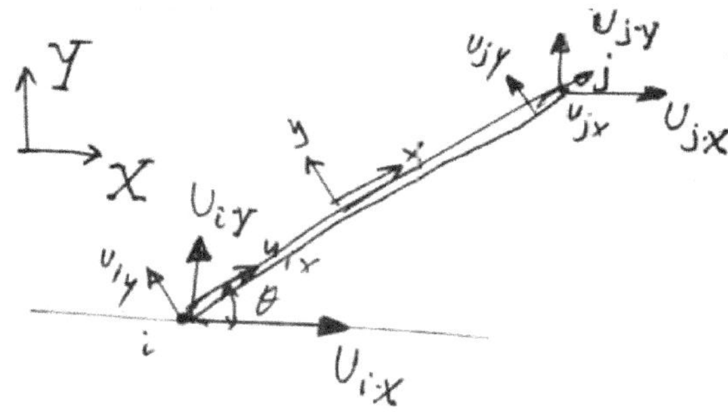


$$U_{iX} = u_{ix}\cos\theta - u_{iy}\sin\theta$$

$$U_{iY} = u_{ix}\sin\theta + u_{iy}\cos\theta$$

$$U_{jX} = u_{jx}\cos\theta - u_{jy}\sin\theta$$

$$U_{jy} = u_{jy}\sin\theta + u_{jx}\cos\theta$$

$\Longrightarrow$ $\boxed{\{U\} = [T]\{u\}}$

where

$$\begin{Bmatrix} U_{iX} \\ U_{iY} \\ \hline U_{jX} \\ U_{jY} \end{Bmatrix} = \underbrace{\left[\begin{array}{cc|cc} \cos\theta & -\sin\theta & 0 & 0 \\ \sin\theta & \cos\theta & 0 & 0 \\ \hline 0 & 0 & \cos\theta & -\sin\theta \\ 0 & 0 & \sin\theta & \cos\theta \end{array}\right]}_{[T]} \begin{Bmatrix} u_{ix} \\ u_{iy} \\ \hline u_{jx} \\ u_{jy} \end{Bmatrix}$$

$[T] \equiv$ Transformation matrix.

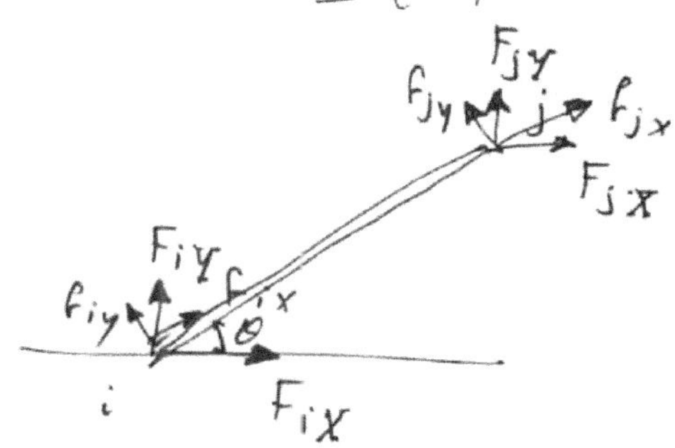


Similarly:

$$F_{iX} = f_{ix}\cos\theta - f_{iy}\sin\theta$$

$$F_{iY} = f_{ix}\sin\theta + f_{iy}\cos\theta$$

$$F_{jX} = f_{jx}\cos\theta - f_{jy}\sin\theta$$

$$F_{jY} = f_{jx}\sin\theta + f_{jy}\cos\theta$$

$\Longrightarrow$ $\boxed{\{F\} = [T]\{f\}}$

where $\{F\} = \begin{Bmatrix} F_{iX} \\ F_{iY} \\ F_{jX} \\ F_{jY} \end{Bmatrix}$, $\{f\} = \begin{Bmatrix} f_{ix} \\ f_{iy} \\ f_{jx} \\ f_{jy} \end{Bmatrix}$

**Note**: the displacements and the forces in the local y-direction are zero.

$$\begin{Bmatrix} f_{ix} \\ f_{jx} \end{Bmatrix} = \begin{bmatrix} k & -k \\ -k & k \end{bmatrix} \begin{Bmatrix} u_{ix} \\ u_{jx} \end{Bmatrix}$$

$$f_{iy} = 0 , \quad f_{jy} = 0$$

$$\Rightarrow \quad \begin{Bmatrix} f_{ix} \\ f_{iy} \\ f_{jx} \\ f_{jy} \end{Bmatrix} = \begin{bmatrix} k & 0 & -k & 0 \\ 0 & 0 & 0 & 0 \\ -k & 0 & k & 0 \\ 0 & 0 & 0 & 0 \end{bmatrix} \begin{Bmatrix} u_{ix} \\ u_{iy} \\ u_{jx} \\ u_{jy} \end{Bmatrix}$$

where $k = \dfrac{EA}{L}$.

$$\Rightarrow \quad \boxed{\{f\} = [k]\{u\}}$$

$\{U\} = [T]\{u\}$ and $\{F\} = [T]\{f\}$

$\{u\} = [T]^{-1}\{U\}$ and $\{f\} = [T]^{-1}\{F\}$

$$\therefore \quad [T]^{-1}\{F\} = [k][T]^{-1}\{U\}$$

$$\Rightarrow \quad \boxed{\{F\} = [T][k][T]^{-1}\{U\}} \quad \leftarrow [k]^{(e)}$$

where
$$[T]^{-1} = \begin{bmatrix} \cos\theta & \sin\theta & 0 & 0 \\ -\sin\theta & \cos\theta & 0 & 0 \\ 0 & 0 & \cos\theta & \sin\theta \\ 0 & 0 & -\sin\theta & \cos\theta \end{bmatrix}$$

$$\therefore \; [k]^{(e)} = [T][k][T]^{-1}$$

$$\therefore \; [k]^{(e)} = k \left[\begin{array}{cc|cc} \cos^2\theta & \sin\theta\cos\theta & -\cos^2\theta & -\sin\theta\cos\theta \\ \sin\theta\cos\theta & \sin^2\theta & -\sin\theta\cos\theta & -\sin^2\theta \\ \hline -\cos^2\theta & -\sin\theta\cos\theta & \cos^2\theta\sin^2\theta & \sin\theta\cos\theta \\ -\sin\theta\cos\theta & -\sin^2\theta & \sin\theta\cos\theta & \sin^2\theta \end{array}\right]$$

---

## Example 2.1 :

page 60

Consider the balcony truss shown in the figure with dimensions. We are interested in determining the deflection of each joint under the loading shown in the figure. All members are made from Douglas-fir wood with a modulus of elasticity of $E = 1.90 \times 10^6 \text{ lb/in}^2$ and a cross-sectional area of $8 \text{ in}^2$. We are also interested in calculating average stresses in each member. Solve this problem manually then use ANSYS.

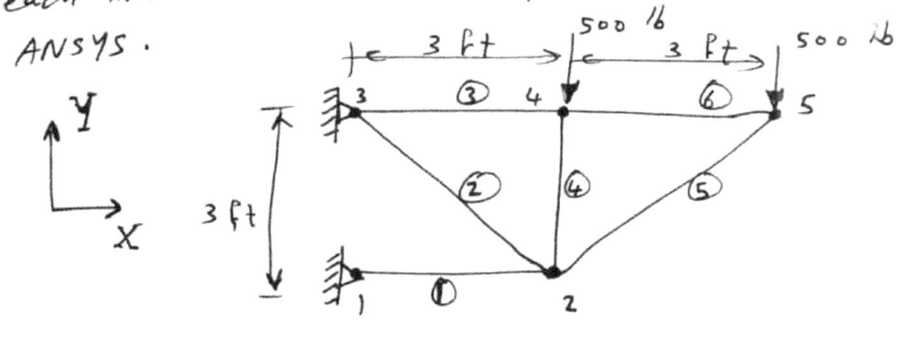


**Solution :** Use 6 elements and 5 nodes.

Determine $\theta$ for each element :

| Element | $\theta$ |
|---|---|
| 1 | 0° |
| 2 | 135° |
| 3 | 0° |
| 4 | 90° |
| 5 | 45° |
| 6 | 0° |

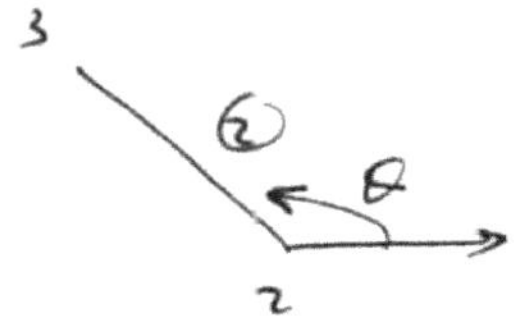

For elements 1, 3, 4, 6 ;

$$k = \frac{EA}{l} = \frac{(1.90\times10^{6})(8)}{3\times12} = 4.22\times10^{5} \text{ lb/in.}$$

For elements 2, 5 :

$$k = \frac{EA}{l} = \frac{(1.90\times10^{6})(8)}{3\sqrt{2}\times12} = 2.98\times10^{5} \text{ lb/in.}$$

Element Stiffness Matrices.

Element ①: $\theta = 0^\circ \Rightarrow \cos\theta = 1,\ \sin\theta = 0.$

$$[k]^{(1)} = 4.22\times10^{5} \begin{array}{c} \begin{array}{cccc} 1x & 1y & 2x & 2y \end{array} \\ \left[\begin{array}{cc|cc} 1 & 0 & -1 & 0 \\ 0 & 0 & 0 & 0 \\ \hline -1 & 0 & 1 & 0 \\ 0 & 0 & 0 & 0 \end{array}\right] \end{array} \begin{array}{c} u_{1x} \\ u_{1y} \\ u_{2x} \\ u_{2y} \end{array} \text{ lb/in.}$$

Element ②: $\theta = 135^\circ \Rightarrow \cos\theta = \frac{-1}{\sqrt{2}},\ \sin\theta = \frac{1}{\sqrt{2}}.$

$$[k]^{(2)} = 2.98\times10^{5} \begin{array}{c} \begin{array}{cccc} 2x & 2y & 3x & 3y \end{array} \\ \left[\begin{array}{cc|cc} \frac{1}{2} & -\frac{1}{2} & -\frac{1}{2} & \frac{1}{2} \\ -\frac{1}{2} & \frac{1}{2} & \frac{1}{2} & -\frac{1}{2} \\ \hline -\frac{1}{2} & \frac{1}{2} & \frac{1}{2} & -\frac{1}{2} \\ \frac{1}{2} & -\frac{1}{2} & -\frac{1}{2} & \frac{1}{2} \end{array}\right] \end{array} \begin{array}{c} u_{2x} \\ u_{2y} \\ u_{3x} \\ u_{3y} \end{array} \text{ lb/in.}$$

Element ③: $\theta = 0^\circ \Rightarrow \cos\theta = 1,\ \sin\theta = 0.$

$$[k]^{(3)} = 4.22\times10^{5} \begin{array}{c} \begin{array}{cccc} 3x & 3y & 4x & 4y \end{array} \\ \left[\begin{array}{cc|cc} 1 & 0 & -1 & 0 \\ 0 & 0 & 0 & 0 \\ \hline -1 & 0 & 1 & 0 \\ 0 & 0 & 0 & 0 \end{array}\right] \end{array} \begin{array}{c} u_{3x} \\ u_{3y} \\ u_{4x} \\ u_{4y} \end{array} \text{ lb/in.}$$

Element ④: $\theta = 90^\circ \Rightarrow \cos\theta = 0,\ \sin\theta = 1$

$$[k]^{(4)} = 4.22\times10^{5} \begin{array}{c} \begin{array}{cccc} 2x & 2y & 4x & 4y \end{array} \\ \left[\begin{array}{cc|cc} 0 & 0 & 0 & 0 \\ 0 & 1 & 0 & -1 \\ \hline 0 & 0 & 0 & 0 \\ 0 & -1 & 0 & 1 \end{array}\right] \end{array} \begin{array}{c} u_{2x} \\ u_{2y} \\ u_{4x} \\ u_{4y} \end{array} \text{ lb/in.}$$

Element ⑤ : $\theta = 45^\circ \Rightarrow \cos\theta = \frac{1}{\sqrt{2}}, \quad \sin\theta = \frac{1}{\sqrt{2}}.$

$$[k]^{(5)} = 2.98 \times 10^5 \begin{array}{c} \begin{array}{cccc} 2x & 2y & 5x & 5y \end{array} \\ \left[\begin{array}{cc|cc} \frac{1}{2} & \frac{1}{2} & -1/2 & -1/2 \\ \frac{1}{2} & \frac{1}{2} & -1/2 & -1/2 \\ \hline -1/2 & -1/2 & \frac{1}{2} & \frac{1}{2} \\ -1/2 & -1/2 & \frac{1}{2} & \frac{1}{2} \end{array}\right] \end{array} \begin{array}{c} \\ u_{2x} \\ u_{2y} \\ u_{5x} \\ u_{5y} \end{array} \quad \text{lb/in.}$$

Element ⑥ : $\theta = 0^\circ \Rightarrow \cos\theta = 1, \quad \sin\theta = 0.$

$$[k]^{(6)} = 4.22 \times 10^5 \begin{array}{c} \begin{array}{cccc} 4x & 4y & 5x & 5y \end{array} \\ \left[\begin{array}{cc|cc} 1 & 0 & -1 & 0 \\ 0 & 0 & 0 & 0 \\ \hline -1 & 0 & 1 & 0 \\ 0 & 0 & 0 & 0 \end{array}\right] \end{array} \begin{array}{c} \\ u_{4x} \\ u_{4y} \\ u_{5x} \\ u_{5y} \end{array} \quad \text{lb/in.}$$

<u>Assembly Process</u>: [K] is a 10x10 matrix.

(<u>Global Stiffness Matrix</u>).

$[K] = 10^5$

| | 1x | 1y | 2x | 2y | 3x | 3y | 4x | 4y | 5x | 5y |
|---|---|---|---|---|---|---|---|---|---|---|
| 1x | 4.22 | | −4.22 | | | | | | | |
| 1y | | | | | | | | | | |
| 2x | −4.22 | | 4.22 +1.49 +1.49 | −1.49 +1.49 | −1.49 | 1.49 | | | −1.49 | −1.49 |
| 2y | | | −1.49 +1.49 | 1.49 +4.22 +1.49 | 1.49 | −1.49 | | −4.22 | −1.49 | −1.49 |
| 3x | | | −1.49 | 1.49 | 1.49 +4.22 | −1.49 | −4.22 | | | |
| 3y | | | 1.49 | −1.49 | −1.49 | 1.49 | | | | |
| 4x | | | | | −4.22 | | 4.22 +4.22 | | −4.22 | |
| 4y | | | | −4.22 | | | | 4.22 | | |
| 5x | | | −1.49 | −1.49 | | | −4.22 | | 1.49 +4.22 | 1.49 |
| 5y | | | −1.49 | −1.49 | | | | | 1.49 | 1.49 |

$$10^5\begin{bmatrix} 4.22 & 0 & -4.22 & 0 & 0 & 0 & 0 & 0 & 0 & 0 \\ 0 & 0 & 0 & 0 & 0 & 0 & 0 & 0 & 0 & 0 \\ \bar{4}.22 & 0 & 7.2 & 0 & -1.49 & 1.49 & 0 & 0 & -1.49 & \bar{1}.49 \\ 0 & 0 & 0 & 7.2 & 1.49 & -1.49 & 0 & \bar{4}.22 & \bar{1}.49 & \bar{1}.49 \\ 0 & 0 & \bar{1}.49 & 1.49 & 5.71 & \bar{1}.49 & \bar{4}.22 & 0 & 0 & 0 \\ 0 & 0 & 1.49 & \bar{1}.49 & \bar{1}.49 & 1.49 & 0 & 0 & 0 & 0 \\ 0 & 0 & 0 & 0 & \bar{4}.22 & 0 & 8.44 & 0 & -4.22 & 0 \\ 0 & 0 & 0 & \bar{4}.22 & 0 & 0 & 0 & 4.22 & 0 & 0 \\ 0 & 0 & \bar{1}.49 & \bar{1}.49 & 0 & 0 & \bar{4}.22 & 0 & 5.71 & 1.49 \\ 0 & 0 & \bar{1}.49 & \bar{1}.49 & 0 & 0 & 0 & 0 & 1.49 & 1.49 \end{bmatrix} \begin{Bmatrix} U_{1x} \; (0) \\ U_{1y} \; (0) \\ U_{2x} \\ U_{2y} \\ U_{3x} \; (0) \\ U_{3y} \; (0) \\ U_{4x} \\ U_{4y} \\ U_{5x} \\ U_{5y} \end{Bmatrix} = \begin{Bmatrix} R_{1x} \\ R_{1y} \\ 0 \\ 0 \\ R_{3x} \\ R_{3y} \\ 0 \\ -500 \\ 0 \\ -500 \end{Bmatrix}$$

Apply boundary conditions:

$$U_{1x} = 0, \quad U_{1y} = 0$$
$$U_{3x} = 0, \quad U_{3y} = 0$$

Solve by partitioning:

$$10^5 \left( -4.22\, U_{2x} \right) = R_{1x} \qquad \text{(1)}$$

$$0 = R_{1y} \qquad \text{(2)}$$

$$10^5\begin{bmatrix} 7.2 & 0 \\ 0 & 7.2 \end{bmatrix}\begin{Bmatrix} U_{2x} \\ U_{2y} \end{Bmatrix} + 10^5\begin{bmatrix} 0 & 0 & \bar{1}.49 & \bar{1}.49 \\ 0 & \bar{4}.22 & \bar{1}.49 & \bar{1}.49 \end{bmatrix}\begin{Bmatrix} U_{4x} \\ U_{4y} \\ U_{5x} \\ U_{5y} \end{Bmatrix} = \begin{Bmatrix} 0 \\ 0 \end{Bmatrix} \qquad \text{(3)}$$

$$10^5\begin{bmatrix} \bar{1}.49 & 1.49 \\ 1.49 & \bar{1}.49 \end{bmatrix}\begin{Bmatrix} U_{2x} \\ U_{2y} \end{Bmatrix} - 10^5\begin{Bmatrix} 4.22 \\ 0 \end{Bmatrix} U_{4x} = \begin{Bmatrix} R_{3x} \\ R_{3y} \end{Bmatrix} \qquad \text{(4)}$$

$$10^5\begin{bmatrix} 0 & 0 \\ 0 & \bar{4}.22 \\ \bar{1}.49 & \bar{1}.49 \\ \bar{1}.49 & \bar{1}.49 \end{bmatrix}\begin{Bmatrix} U_{2x} \\ U_{2y} \end{Bmatrix} + 10^5\begin{bmatrix} 8.44 & 0 & \bar{4}.22 & 0 \\ 0 & 4.22 & 0 & 0 \\ \bar{4}.22 & 0 & 5.71 & 1.49 \\ 0 & 0 & 1.49 & 1.49 \end{bmatrix}\begin{Bmatrix} U_{4x} \\ U_{4y} \\ U_{5x} \\ U_{5y} \end{Bmatrix} = \begin{Bmatrix} 0 \\ -500 \\ 0 \\ -500 \end{Bmatrix} \quad (5)$$

From eq. (3):

$$\begin{bmatrix} 7.2 & 0 \\ 0 & 7.2 \end{bmatrix}\begin{Bmatrix} U_{2x} \\ U_{2y} \end{Bmatrix} = \begin{bmatrix} 0 & 0 & 1.49 & 1.49 \\ 0 & 4.22 & 1.49 & 1.49 \end{bmatrix}\begin{Bmatrix} U_{4x} \\ U_{4y} \\ U_{5x} \\ U_{5y} \end{Bmatrix}$$

$$\therefore \begin{Bmatrix} U_{2x} \\ U_{2y} \end{Bmatrix} = \underset{2\times2}{\begin{bmatrix} \frac{1}{7.2} & 0 \\ 0 & \frac{1}{7.2} \end{bmatrix}}\underset{2\times4}{\begin{bmatrix} 0 & 0 & 1.49 & 1.49 \\ 0 & 4.22 & 1.49 & 1.49 \end{bmatrix}}\begin{Bmatrix} U_{4x} \\ U_{4y} \\ U_{5x} \\ U_{5y} \end{Bmatrix}$$

$$\therefore \begin{Bmatrix} U_{2x} \\ U_{2y} \end{Bmatrix} = \begin{bmatrix} 0 & 0 & 0.207 & 0.207 \\ 0 & 0.586 & 0.207 & 0.207 \end{bmatrix}\begin{Bmatrix} U_{4x} \\ U_{4y} \\ U_{5x} \\ U_{5y} \end{Bmatrix}$$

Substitute the above into eq. (5) to obtain:

$$10^5\underset{4\times2}{\begin{bmatrix} 0 & 0 \\ 0 & \bar{4}.22 \\ \bar{1}.49 & \bar{1}.49 \\ \bar{1}.49 & \bar{1}.49 \end{bmatrix}}\begin{bmatrix} 0 & 0 & 0.207 & 0.207 \\ 0 & 0.586 & 0.207 & 0.207 \end{bmatrix}\begin{Bmatrix} U_{4x} \\ U_{4y} \\ U_{5x} \\ U_{5y} \end{Bmatrix} + 10^5 \cdots = \begin{Bmatrix} 0 \\ -500 \\ 0 \\ -500 \end{Bmatrix}$$

$$\therefore 10^5\left(\begin{bmatrix} 0 & 0 & 0 & 0 \\ 0 & \bar{2}.47 & \bar{0}.874 & \bar{0}.874 \\ 0 & \bar{0}.874 & \bar{0}.617 & \bar{0}.617 \\ 0 & \bar{0}.874 & \bar{0}.617 & \bar{0}.617 \end{bmatrix} + \begin{bmatrix} 8.44 & 0 & \bar{4}.22 & 0 \\ 0 & 4.22 & 0 & 0 \\ \bar{4}.22 & 0 & 5.71 & 1.49 \\ 0 & 0 & 1.49 & 1.49 \end{bmatrix}\right)\begin{Bmatrix} U_{4x} \\ U_{4y} \\ U_{5x} \\ U_{5y} \end{Bmatrix} = \begin{Bmatrix} 0 \\ -500 \\ 0 \\ -500 \end{Bmatrix}$$

$$10^5 \begin{bmatrix} 8.44 & 0 & -4.22 & 0 \\ 0 & 1.75 & -0.874 & -0.874 \\ -4.22 & -0.874 & 5.09 & 0.873 \\ 0 & -0.874 & 0.873 & 0.873 \end{bmatrix} \begin{Bmatrix} U_{4x} \\ U_{4y} \\ U_{5x} \\ U_{5y} \end{Bmatrix} = \begin{Bmatrix} 0 \\ -500 \\ 0 \\ -500 \end{Bmatrix}$$

Solve by Gaussian elimination to obtain:

$$\begin{Bmatrix} U_{4x} \\ U_{4y} \\ U_{5x} \\ U_{5y} \end{Bmatrix} = \begin{Bmatrix} 0.00118 \\ -0.0114 \\ 0.00240 \\ -0.0195 \end{Bmatrix} \text{ in.}$$

then

$$\begin{Bmatrix} U_{2x} \\ U_{2y} \end{Bmatrix} = \begin{Bmatrix} -0.00355 \\ -0.01026 \end{Bmatrix} \text{ in.}$$

and

$$R_{1x} = 1500 \text{ lb.}$$
$$R_{3x} = -1500 \text{ lb.}$$
$$R_{3y} = 1000 \text{ lb.}$$

To find the reactions systematically, use:

$$\boxed{\{R\} = [K]_{10\times10}\{U\}_{10\times1} - \{F\}_{10\times1}}$$

where

$$\{U\} = \begin{Bmatrix} 0 \\ 0 \\ -0.00355 \\ -0.01026 \\ 0 \\ 0 \\ 0.00118 \\ -0.0114 \\ 0.0024 \\ -0.0195 \end{Bmatrix}_{10\times1} \text{ in.}, \quad \{F\} = \begin{Bmatrix} 0 \\ 0 \\ 0 \\ 0 \\ 0 \\ 0 \\ 0 \\ -500 \\ 0 \\ -500 \end{Bmatrix} \text{ lb.}$$

## Element stresses:

Internal forces:

$$f_{ix} = k(u_{ix} - u_{jx})$$

$$f_{jx} = k(u_{jx} - u_{ix})$$

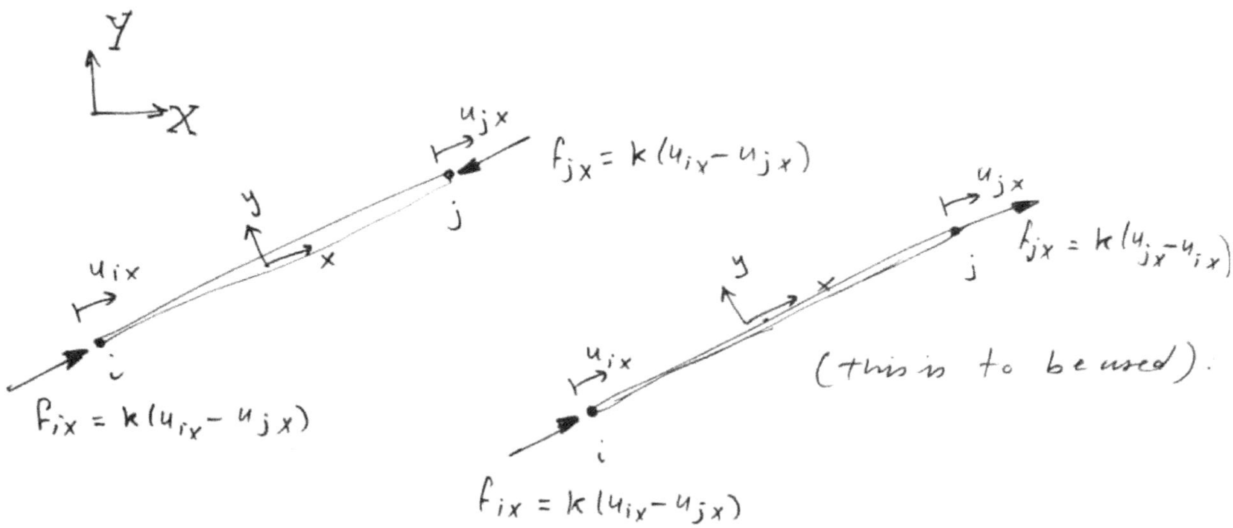


We need the local displacements $u_{ix}, u_{jx}$:

$$\{U\} = [T]\{u\}$$

$$\Rightarrow \quad \{u\} = [T]^{-1}\{U\}$$

$$\therefore \quad \begin{Bmatrix} u_{ix} \\ u_{iy} \\ \hline u_{jx} \\ u_{jy} \end{Bmatrix} = \left[\begin{array}{cc|cc} \cos\theta & \sin\theta & 0 & 0 \\ -\sin\theta & \cos\theta & 0 & 0 \\ \hline 0 & 0 & \cos\theta & \sin\theta \\ 0 & 0 & -\sin\theta & \cos\theta \end{array}\right] \begin{Bmatrix} U_{ix} \\ U_{iy} \\ U_{jx} \\ U_{jy} \end{Bmatrix}$$

Average normal stresses: $\sigma = \dfrac{f}{A}$.

$$\sigma = \frac{f}{A} = \frac{k(u_{ix} - u_{jx})}{A} = \frac{\frac{EA}{l}(u_{ix} - u_{jx})}{A} = \frac{E(u_{ix} - u_{jx})}{l}.$$

Compute the normal stress in element ⑤ :

Element ⑤ : $\theta = 45^\circ$ .

$$U_{2x} = -0.00355 \text{ in.}$$
$$U_{2y} = -0.01026 \text{ in.}$$
$$U_{5x} = 0.0024 \text{ in.}$$
$$U_{5y} = -0.0195 \text{ in.}$$

$$\therefore \begin{Bmatrix} u_{2x} \\ u_{2y} \\ u_{5x} \\ u_{5y} \end{Bmatrix} = \begin{bmatrix} \cos 45^\circ & \sin 45^\circ & 0 & 0 \\ -\sin 45^\circ & \cos 45^\circ & 0 & 0 \\ 0 & 0 & \cos 45^\circ & \sin 45^\circ \\ 0 & 0 & -\sin 45^\circ & \cos 45^\circ \end{bmatrix} \begin{Bmatrix} -0.00355 \\ -0.01026 \\ 0.0024 \\ -0.0195 \end{Bmatrix}$$

$$\therefore \begin{Bmatrix} u_{2x} \\ u_{2y} \\ u_{5x} \\ u_{5y} \end{Bmatrix} = \begin{Bmatrix} -0.009765 \\ -0.004745 \\ -0.01209 \\ -0.01549 \end{Bmatrix} \text{ in.}$$

$$\Rightarrow f_{2x} = k(u_{2x} - u_{5x})$$
$$= \frac{EA}{l}(u_{2x} - u_{5x})$$
$$= \frac{(1.90\times10^6)(8)}{3\sqrt{2}\times12}(-0.009765 - (-0.01209))$$
$$= 694 \text{ lb.}$$

$$\sigma_5 = \frac{E(u_{2x} - u_{5x})}{l} = \frac{694}{A} = \frac{694}{8} = 87 \text{ lb/in}^2.$$

## 2.3 Space Trusses :

Page 71

space truss ≡ three-dimensional truss.

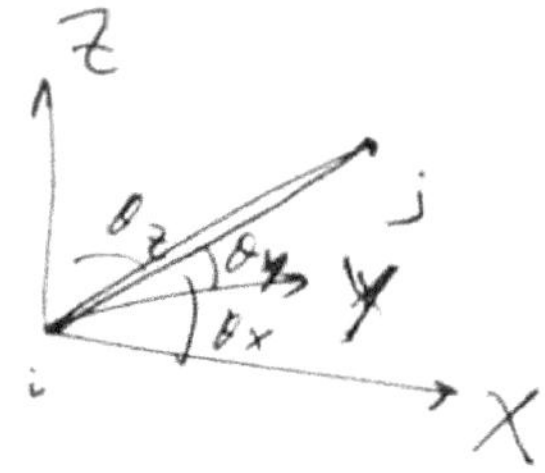


$$\{U\} = \begin{Bmatrix} U_{ix} \\ U_{iy} \\ U_{iz} \\ \hline U_{jx} \\ U_{jy} \\ U_{jz} \end{Bmatrix}$$

$$\cos\theta_x = \frac{X_j - X_i}{L}, \quad \cos\theta_y = \frac{Y_j - Y_i}{L}, \quad \cos\theta_z = \frac{Z_j - Z_i}{L}$$

where $L = \sqrt{(X_j - X_i)^2 + (Y_j - Y_i)^2 + (Z_j - Z_i)^2}$

$[K]^{(e)}$ for each element is a 6X6 matrix. (see page 73).

## 2.4 Overview of the ANSYS Program :

Page 73

Two lab sessions :

Lab Session 1 : Plane Truss (pages 80 – 95).

Lab Session 2 : Space Truss (pages 96 – 109).

## 2.6 Verification of Results: page 110 (4)

1 - Check the reactions.

2 - Sum of all forces at each node should be zero.

3 - Pass an arbitrary section through the truss.

(see pages 110 – 111).

---

**Homework 2** page 113, Problems 3 & 4.

---

**Homework 3** ANSYS problems: problem 9 page 115.

problem 20 page 120. (Design problem).

# Chapter 3

## One-Dimensional Elements

page 121

### 3.1 Linear Elements:

page 121

We may approximate the actual deflection of the column by using a series of linear functions, describing the deflection over each element.

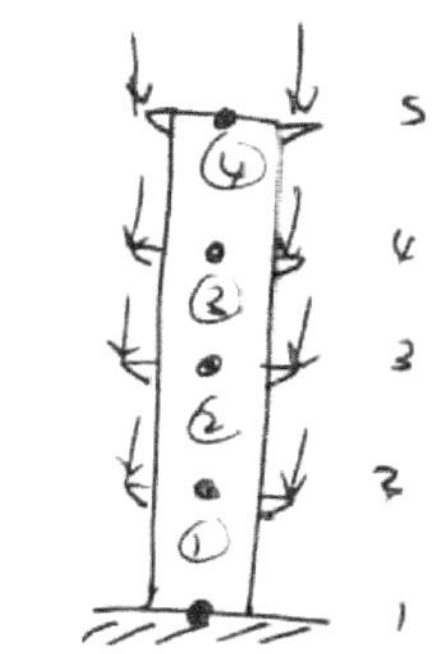


Axially loaded elements.

For a typical element, the displacement distribution is given by:

$$u^{(e)} = c_1 + c_2 X \qquad \text{(linear function)}.$$

at $X = X_i$, $u = u_i$

$X = X_j$, $u = u_j$

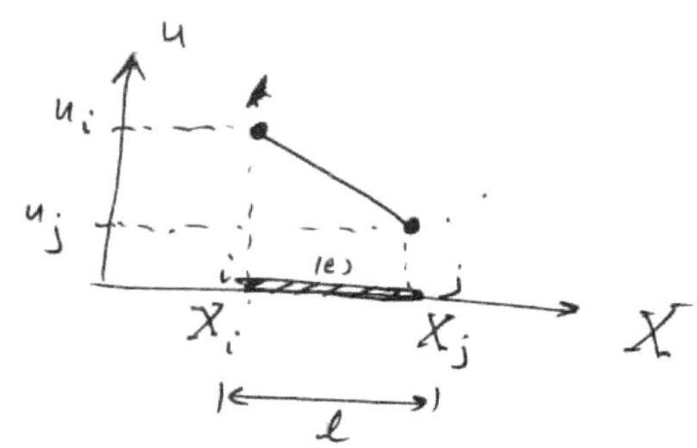


$$\begin{cases} u_i = c_1 + c_2 X_i & \text{--- (1)} \\ u_j = c_1 + c_2 X_j & \text{--- (2)} \end{cases}$$

subtract:

$$u_i - u_j = c_2 (X_i - X_j)$$

$$\Rightarrow \quad c_2 = \frac{u_j - u_i}{X_j - X_i}$$

$$\Rightarrow \quad c_1 = u_i - c_2 X_i = u_i - \left(\frac{u_j - u_i}{X_j - X_i}\right) X_i$$

$$= \frac{u_i (X_j - X_i) - X_i (u_j - u_i)}{X_j - X_i}$$

$$\therefore c_1 = \frac{u_i X_j - u_j X_i}{X_j - X_i}$$

$$\Rightarrow u^{(e)} = c_1 + c_2 X$$

$$u^{(e)} = \frac{u_i X_j - u_j X_i}{X_j - X_i} + \frac{u_j - u_i}{X_j - X_i} X$$

re-arrange the terms:

$$u^{(e)} = \frac{(X_j - X) u_i + (X - X_i) u_j}{X_j - X_i}$$

$$\Rightarrow u^{(e)} = \left(\frac{X_j - X}{X_j - X_i}\right) u_i + \left(\frac{X - X_i}{X_j - X_i}\right) u_j$$

Define
$$S_i = \frac{X_j - X}{X_j - X_i} = \frac{X_j - X}{l}, \quad l = X_j - X_i.$$
$$S_j = \frac{X - X_i}{X_j - X_i} = \frac{X - X_i}{l}$$

→ **shape functions** or **interpolation functions.**

$$\Rightarrow u^{(e)} = S_i u_i + S_j u_j$$

$u_i, u_j$: deflections of nodes $i, j$.

$$\boxed{u^{(e)} = [S_i \quad S_j] \begin{Bmatrix} u_i \\ u_j \end{Bmatrix}}$$

In general, for any unknown $\psi$:

$$\psi^{(e)} = [S_i \quad S_j] \begin{Bmatrix} \psi_i \\ \psi_j \end{Bmatrix}$$

## Properties of Shape Functions:

$$S_i = \frac{X_j - X}{\ell} \quad , \quad S_j = \frac{X - X_i}{\ell}$$

(1)

$$S_i\Big|_{X=X_i} = \frac{X_j - X_i}{\ell} = 1 \quad , \quad S_i\Big|_{X=X_j} = \frac{X_j - X_j}{\ell} = 0 .$$

$$S_j\Big|_{X=X_i} = \frac{X_i - X_i}{\ell} = 0 \quad , \quad S_j\Big|_{X=X_j} = \frac{X_j - X_i}{\ell} = 1 .$$

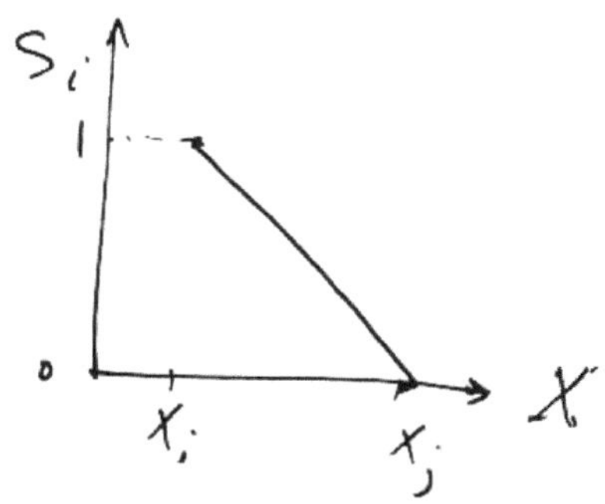


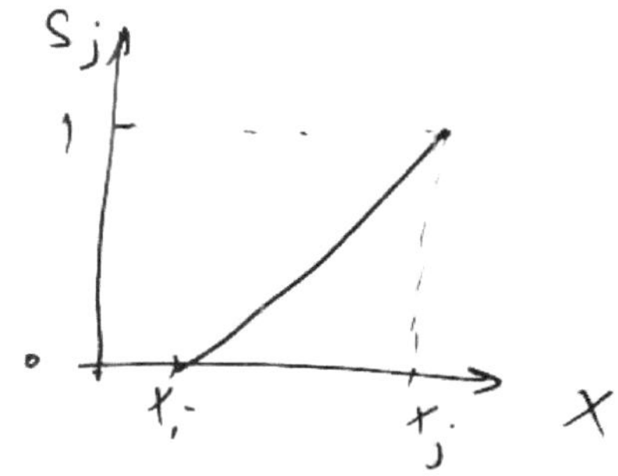


Sum of Shape Functions:

(2) $$S_i + S_j = \frac{X_j - X}{\ell} + \frac{X - X_i}{\ell} = 1 .$$

$\longrightarrow$ $\boxed{S_i + S_j = 1}$ always.

(3) Sum of Derivatives of Shape Functions:

$$\frac{d}{dX}(S_i) + \frac{d}{dX}(S_j) = -\frac{1}{\ell} + \frac{1}{\ell} = 0 .$$

---

page 125

Example 3.1 : omitted.
temperature

---

## Example 3.2 :

Consider a four-story building with steel columns. One column is subjected to the loading shown in the figure. Under axial loading assumption and using linear elements, the vertical displacements of the column at various floor-column connection points were determined to be:

$$\begin{Bmatrix} u_1 \\ u_2 \\ u_3 \\ u_4 \\ u_5 \end{Bmatrix} = \begin{Bmatrix} 0 \\ 0.03283 \\ 0.05784 \\ 0.07504 \\ 0.08442 \end{Bmatrix} \text{ in.}$$

the modulus of elasticity of $E = 29 \times 10^6$ lb/in$^2$ and area of $A = 39.7$ in$^2$ were used in the calculations. A detailed analysis of this problem is given in Chapter 4. Determine the deflections at points A and B.

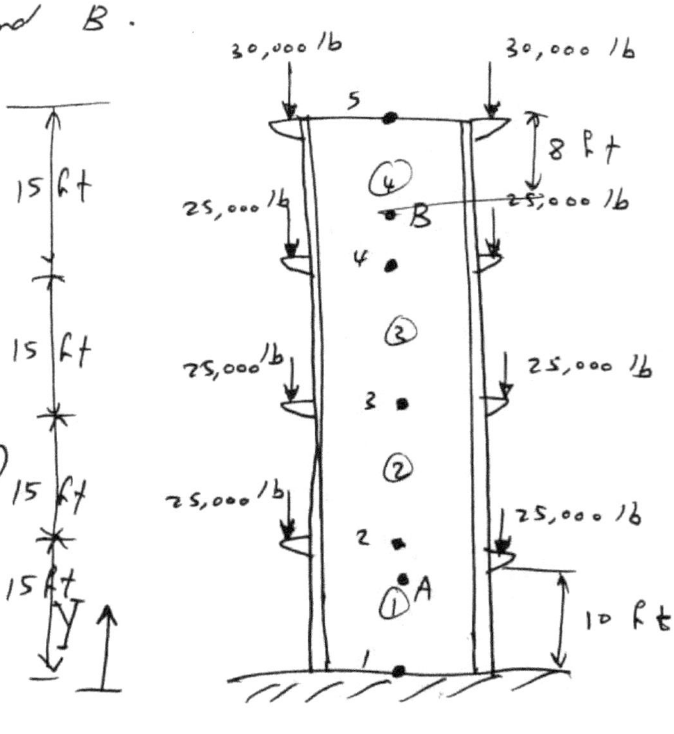


Solution :

Use $Y$ instead of $X$ :

(a) Point A :

$$u_A^{(1)} = S_1^{(1)} u_1 + S_2^{(1)} u_2$$

$$= \frac{Y_2 - Y}{\ell} u_1 + \frac{Y - Y_1}{\ell} u_2$$

$$= \frac{15 - 10}{15}(0) + \frac{10 - 0}{15}(0.03283)$$

$$u_A^{(1)} = 0.02189 \text{ in.}$$

(b) Point B:

$$u_B^{(4)} = S_4^{(4)} u_4 + S_5^{(4)} u_5$$

$$= \frac{Y_5 - Y}{\ell} u_4 + \frac{Y - Y_4}{\ell} u_5$$

$$= \frac{60-52}{15}(0.07504) + \frac{52-45}{15}(0.08442)$$

$$u_B^{(4)} = 0.07942 \text{ in.}$$

---

## 3.2 Quadratic Elements:

page 127

Increase accuracy by either:

1- increasing the number of elements used.

2- using higher-order shape functions for the elements.

Using a quadratic function, we need three nodes/element.

$$u^{(e)} = c_1 + c_2 X + c_3 X^2$$

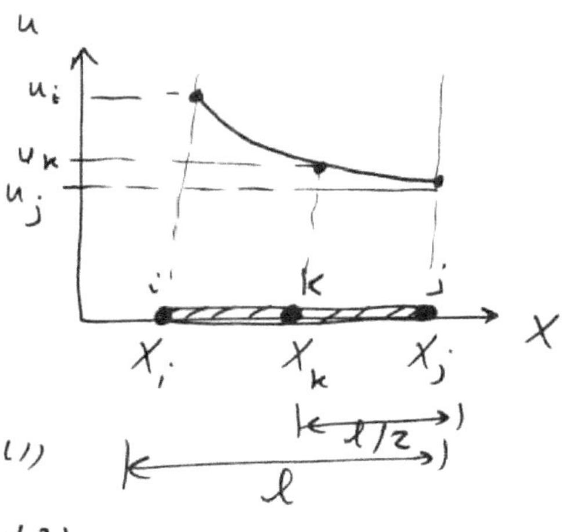


at $X = X_i, \quad u = u_i$

$X = X_j, \quad u = u_j$

$X = X_k, \quad u = u_k$

$$\Rightarrow u_i = c_1 + c_2 X_i + c_3 X_i^2 \quad \text{——} (1)$$

$$u_j = c_1 + c_2 X_j + c_3 X_j^2 \quad \text{——} (2)$$

$$u_k = c_1 + c_2 X_k + c_3 X_k^2 \quad \text{——} (3)$$

Solve for $c_1, c_2, c_3$ and re-arrange the terms:

$$u^{(e)} = S_i u_i + S_j u_j + S_k u_k$$

$$u^{(e)} = [S_i \quad S_j \quad S_k] \begin{Bmatrix} u_i \\ u_j \\ u_k \end{Bmatrix}$$

where the quadratic shape functions are:

$$S_i = \frac{2}{\ell^2}(X - X_j)(X - X_k)$$

$$S_j = \frac{2}{\ell^2}(X - X_i)(X - X_k)$$

$$S_k = -\frac{4}{\ell^2}(X - X_i)(X - X_j)$$

<u>properties of shape functions:</u>

1 - A shape function has a value of 1 at its corresponding node and a value of 0 at the other adjacent node.

2 - $S_i + S_j + S_k = 1$.

3 - $\frac{d}{dX}(S_i) + \frac{d}{dX}(S_j) + \frac{d}{dX}(S_k) \neq 0$.

---

page 128

## 3.3 Cubic Elements:

Use four-node elements.

Let $u^{(e)} = c_1 + c_2 X + c_3 X^2 + c_4 X^3$

$$u_i = c_1 + c_2 X_i + c_3 X_i^2 + c_4 X_i^3$$

$$u_j = c_1 + c_2 X_j + c_3 X_j^2 + c_4 X_j^3$$

$$u_k = c_1 + c_2 X_k + c_3 X_k^2 + c_4 X_k^3$$

$$u_m = c_1 + c_2 X_m + c_3 X_m^2 + c_4 X_m^3$$

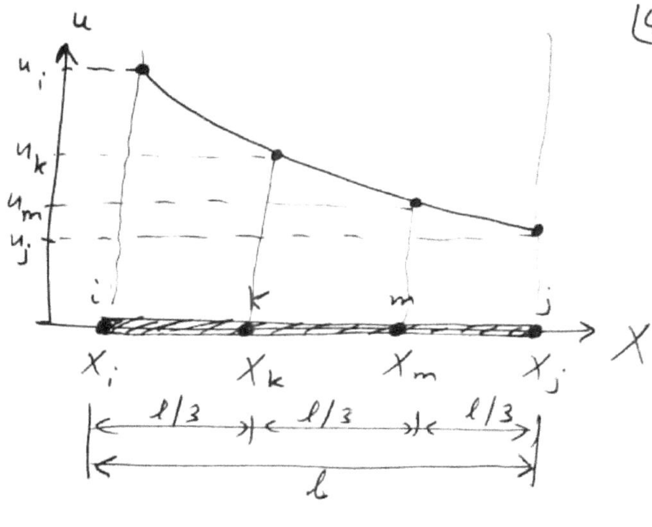


$\Longrightarrow$

$$u^{(e)} = S_i T_i + S_j T_j + S_k T_k + S_m T_m$$

$$\Rightarrow \quad u^{(e)} = [S_i \quad S_j \quad S_k \quad S_m] \begin{Bmatrix} u_i \\ u_j \\ u_k \\ u_m \end{Bmatrix}$$

$\Rightarrow$ the cubic shape functions are:

$$S_i = -\frac{9}{2\ell^3}(X - X_j)(X - X_k)(X - X_m)$$

$$S_j = \frac{9}{2\ell^3}(X - X_i)(X - X_k)(X - X_m)$$

$$S_k = \frac{27}{2\ell^3}(X - X_i)(X - X_j)(X - X_m)$$

$$S_m = -\frac{27}{2\ell^3}(X - X_i)(X - X_j)(X - X_k)$$

---

Instead of solving simultaneous equations, we can use "Lagrange interpolation functions"

Let $S_i = a_1 (X - X_j)(X - X_k)(X - X_m)$

should satisfy $S_i = 1$ at $X = X_i$

$S_i = 0$ at $X = X_j, X_k, X_m$

$$\therefore \quad 1 = a_1 (X_i - X_j)(X_i - X_k)(X_i - X_m)$$

$$1 = a_1 (-\ell)(-\tfrac{\ell}{3})(-2\tfrac{\ell}{3})$$

$$\Rightarrow a_1 = -\frac{9}{2\ell^3}.$$

$$\Rightarrow S_i = -\frac{9}{2\ell^3}(X - X_j)(X - X_k)(X - X_m)$$

Similarly, define $S_j, S_k, S_m$.

---

## 3.4 Global, Local, and Natural Coordinates: page 131

$$\boxed{X = X_i + x}$$

where $X$ : global coordinate.

$x$ : local coordinate.

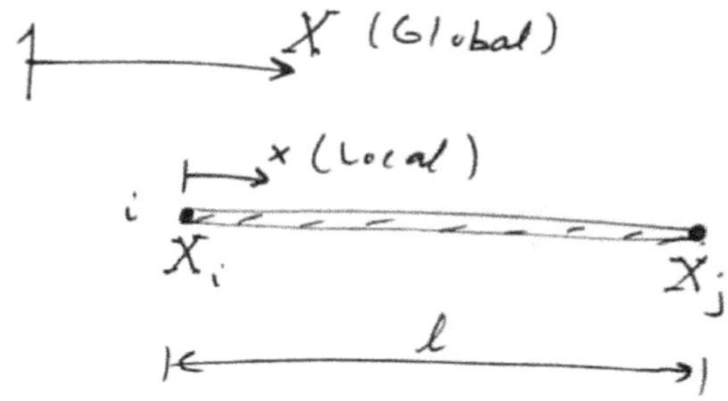


$$\Rightarrow X_i = X - x$$

$$X_j = X_i + \ell = X - x + \ell.$$

$$\therefore S_i = \frac{X_j - X}{\ell} = \frac{(X - x + \ell) - X}{\ell} = \frac{\ell - x}{\ell} = 1 - \frac{x}{\ell}.$$

$$S_j = \frac{X - X_i}{\ell} = \frac{X - (X - x)}{\ell} = \frac{x}{\ell}.$$

$$\Longrightarrow \left.\begin{array}{l} S_i = 1 - \frac{x}{\ell} \\ S_j = \frac{x}{\ell} \end{array}\right\} \text{ linear shape functions, where } 0 \le x \le \ell$$

## One-Dimensional Linear Natural Coordinates.

* Natural coordinates are local coordinates in a dimensionless form.

let $\boxed{\xi = \frac{2x}{\ell} - 1}$ natural coordinate.

at $x = 0$, $\xi = -1$ (node i)
$x = \ell$, $\xi = +1$ (node j)

$$2x - \ell = \xi\ell$$
$$2x = \xi\ell + \ell$$
$$x = \frac{\ell(1+\xi)}{2}.$$

∴ Natural linear shape functions:

$$S_i = 1 - \frac{x}{\ell} = 1 - \frac{1}{\ell}\cdot\frac{\ell}{2}(1+\xi)$$
$$= 1 - \frac{1}{2} - \frac{1}{2}\xi = \frac{1}{2} - \frac{1}{2}\xi$$
$$S_i = \frac{1}{2}(1-\xi)$$
$$S_j = \frac{x}{\ell} = \frac{1}{\ell}\cdot\frac{\ell}{2}(1+\xi) = \frac{1}{2}(1+\xi).$$

$$\Longrightarrow \boxed{\begin{aligned} S_i &= \tfrac{1}{2}(1-\xi) \\ S_j &= \tfrac{1}{2}(1+\xi) \end{aligned}}$$

**Displacement Field:**

$$u^{(e)} = S_i u_i + S_j u_j = \frac{1}{2}(1-\xi)u_i + \frac{1}{2}(1+\xi)u_j$$

At node i, $\xi = -1$, $u^{(e)} = u_i$.

At node j, $\xi = +1$, $u^{(e)} = u_j$.

**Displacement Field:**

$$X = S_i X_i + S_j X_j = \frac{1}{2}(1-\xi)X_i + \frac{1}{2}(1+\xi)X_j \quad \text{(Global)}$$

$$x = S_i x_i + S_j x_j = \frac{1}{2}(1-\xi)x_i + \frac{1}{2}(1+\xi)x_j \quad \text{(Local)}.$$

**Note**: We used the same shape functions $S_i, S_j$ for the displacement $u$ and the distance $X$. This is called an *isoparametric* finite element formulation.

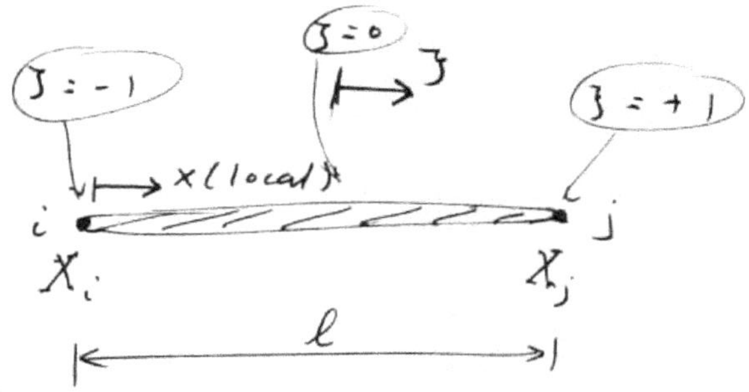


page 133

**Example 3.3 (omitted).**

**One-Dimensional Natural Quadratic and Cubic Shape Functions.**

Quadratic:

$$S_i = -\tfrac{1}{2}\xi(1-\xi)$$
$$S_j = \tfrac{1}{2}\xi(1+\xi)$$
$$S_k = (1+\xi)(1-\xi)$$

Cubic:

$$S_i = \tfrac{1}{16}(1-\xi)(3\xi+1)(3\xi-1)$$
$$S_j = \tfrac{1}{16}(1+\xi)(3\xi+1)(3\xi-1)$$
$$S_k = \tfrac{9}{16}(1+\xi)(\xi-1)(3\xi-1)$$
$$S_m = \tfrac{9}{16}(1+\xi)(1-\xi)(3\xi+1)$$

See Table page 135 for summary.

page 134

**Example 3.4:**

Evaluate the integral $\int_{X_i}^{X_j} S_j^2\, dX$ using (a) global coordinates. (b) local coordinates.

**Solution:**

(a)
$$\int_{X_i}^{X_j} S_j^2\, dX = \int_{X_i}^{X_j} \left(\frac{X-X_i}{\ell}\right)^2 dX = \left.\frac{(X-X_i)^3}{3\ell^2}\right|_{X_i}^{X_j}$$

$$= \frac{(X_j - X_i)^3}{3\ell^2} = \frac{\ell^3}{3\ell^2} = \frac{\ell}{3}.$$

(b) $\int_{X_i}^{X_j} S_j^2 \, dX = \int_0^{\ell} \left(\frac{x}{\ell}\right)^2 dx$ , note $dX = dx$

$= \frac{x^3}{3\ell^2}\Big]_0^{\ell} = \frac{\ell^3}{3\ell^2} = \frac{\ell}{3}$ .

---

## 3.5 Numerical Integration : Gauss–Legendre Quadrature :

page 136

Use the two-point Gauss–Legendre formula:

$$I = \int_a^b f(x)dx = \sum_{i=1}^{n} w_i f(x_i) ,$$

How do we select the weights $w_i$ and points $x_i$ ?

Change the limits of integration from a, b to −1, +1 :

$x = c_0 + c_1 \lambda$  ( introduce the variable $\lambda$ ) :

add
$$\begin{cases} a = c_0 + c_1(-1) \\ b = c_0 + c_1(+1) \end{cases}$$
solve for $c_0, c_1$ :

$a + b = 2c_0 \Rightarrow c_0 = \frac{a+b}{2} = \frac{b+a}{2}$ .

$c_1 = c_0 - a = \frac{b+a}{2} - a = \frac{b-a}{2}$ .

$\therefore \quad x = \left(\frac{b+a}{2}\right) + \left(\frac{b-a}{2}\right)\lambda$

$dx = \left(\frac{b-a}{2}\right) d\lambda$

$\therefore \quad I = \int_{-1}^{+1} f(\lambda) d\lambda = \sum_{i=1}^{n} w_i f(\lambda_i)$

## The Two-Point Gauss-Legendre Formula:

Use two weights $w_1, w_2$ and two sampling points $\lambda_1, \lambda_2$:

$$\int_{-1}^{+1} f(\lambda)\,d\lambda = \sum_{i=1}^{n} w_i f(\lambda_i) \quad , \quad \text{use } \boxed{n=2} \;.$$

$$= w_1 f(\lambda_1) + w_2 f(\lambda_2)$$

Use the Legendre polynomials

$f(\lambda) = 1$
$f(\lambda) = \lambda$
$f(\lambda) = \lambda^2$
$f(\lambda) = \lambda^3$

} four polynomials

$$\int_{-1}^{+1} 1\,d\lambda = \lambda\Big]_{-1}^{+1} = 1-(-1) = \boxed{2 = w_1 f(\lambda_1) + w_2 f(\lambda_2)} \text{ —— (1)}$$

$$\int_{-1}^{+1} \lambda\,d\lambda = \frac{\lambda^2}{2}\Big]_{-1}^{+1} = \boxed{0 = w_1 f(\lambda_1) + w_2 f(\lambda_2)} \text{ —— (2)}$$

$$\int_{-1}^{+1} \lambda^2\,d\lambda = \frac{\lambda^3}{3}\Big]_{-1}^{+1} = \boxed{\frac{2}{3} = w_1 f(\lambda_1) + w_2 f(\lambda_2)} \text{ —— (3)}$$

$$\int_{-1}^{+1} \lambda^3\,d\lambda = \frac{\lambda^4}{4}\Big]_{-1}^{+1} = \boxed{0 = w_1 f(\lambda_1) + w_2 f(\lambda_2)} \text{ —— (4)}$$

---

Substitute:

$$\begin{cases} w_1 + w_2 = 2 & \text{—— (1)} \\ w_1\lambda_1 + w_2\lambda_2 = 0 & \text{—— (2)} \\ w_1\lambda_1^2 + w_2\lambda_2^2 = \frac{2}{3} & \text{—— (3)} \\ w_1\lambda_1^3 + w_2\lambda_2^3 = 0 & \text{—— (4)} \end{cases}$$

Solve equations (1), (2), (3) & (4) simultaneously:

$w_1 = 1$

$w_2 = 1$

$\lambda_1 = -0.577350269$

$\lambda_2 = 0.557735026 9$

The accuracy increases as the number of sampling points increases.

See Table page 138 for different weights and sampling points.

---

Example 3.5: page 137

Evaluate the integral $I = \int_2^6 (x^2 + 5x + 3)dx$ using the Gauss-Legendre two-point sampling formula.

Solution:

Exact Solution: $\int_2^6 (x^2 + 5x + 3)dx = \left[\frac{x^3}{3} + 5\frac{x^2}{2} + 3x\right]_2^6$

$= 180 - 18.667$

$= 161.33333333$ (Exact).

Use the two-point Gauss-Legendre formula:

$$x = \left(\frac{b+a}{2}\right) + \left(\frac{b-a}{2}\right)\lambda = \left(\frac{6+2}{2}\right) + \left(\frac{6-2}{2}\right)\lambda$$

$\therefore$

$$x = 4 + 2\lambda$$

$$dx = 2d\lambda$$

$$\therefore \quad I = \int_2^6 (x^2 + 5x + 3)dx = \int_{-1}^{+1} \left[(4+2\lambda)^2 + 5(4+2\lambda) + 3\right] 2d\lambda$$

$$I \cong w_1 f(\lambda_1) + w_2 f(\lambda_2)$$

$$= 1 \cdot f(-0.577350269) + 1 f(+0.577350269)$$

$$= 50.64445268 + 110.6888807$$

$$= 161.333333333$$

---

Example 3.6 :

Evaluate the integral $\int_{X_i}^{X_j} S_j^2 \, dX$ using the Gauss-Legendre two-point formula ?

Solution : $S_j = \frac{1}{2}(1+\xi)$ , $dx = \frac{\ell}{2} d\xi$ , here use $\xi = \lambda$ .

$$\therefore I = \int_{X_i}^{X_j} S_j^2 \, dX = \int_{X_i}^{X_j} \left(\frac{X - X_i}{\ell}\right)^2 dX = \int_0^{\ell} \left(\frac{x}{\ell}\right)^2 dx = \frac{\ell}{2} \int_{-1}^{+1} \left[\frac{1}{2}(1+\xi)\right]^2 d\xi$$

$$\therefore I = w_1 f(\lambda_1) + w_2 f(\lambda_2)$$

($\lambda_1 \leftarrow \xi_1$, $\lambda_2 \leftarrow \xi_2$)

$$= 1 \, f(-0.577350269) + 1 \, f(+0.577350269)$$

$$= 0.022329099 \, \ell + 0.311004233 \, \ell$$

$$= 0.33333333 \, \ell = \frac{\ell}{3} .$$

---

3.6 Examples of One-Dimensional Elements in ANSYS :

page 139

uniaxial link elements :

LINK31 : uniaxial heat conduction element.

LINK32 : for radiation heat transfer.

LINK34 : uniaxial connection link.

---

Homework 4 ) pages 140 – 144 :

problems 2 , 4 , 12 , 19 , 21 .

NOT REQUIRED

# Chapter 4

## Analysis of One-Dimensional Problems

page 145

## 4.1 Heat Transfer Problems: (omitted).

## 4.2 Solid Mechanics Problems:

page 162

For a column (axial-load member).

use $$[k]^{(e)} = \frac{EA}{\ell}\begin{bmatrix} 1 & -1 \\ -1 & 1 \end{bmatrix}$$

Derivation pages 162/163 is not required.

---

## Example 4.4:

page 164

Consider four-story building with steel columns. One column is subjected to the loading shown in the figure. Assuming axial loading, determine:

(a) vertical displacements of the column at various floor-column connection points.

(b) stresses in each portion of the column.

Use $E = 29 \times 10^6$ lb/in$^2$ and $A = 39.7$ in$^2$.

**Solution:**

For each portion:

$$\frac{EA}{\ell} = \frac{(29\times10^6)(39.7)}{15\times12} = 6.396\times10^6 \text{ lb/in.}$$

$$\therefore [k]^{(1)} = 6.396\times10^6 \begin{bmatrix} 1 & -1 \\ -1 & 1 \end{bmatrix} \begin{matrix} 1 \\ 2 \end{matrix} \text{ lb/in.} \quad (\text{columns } 1, 2)$$

$$[k]^{(2)} = 6.396\times10^6 \begin{bmatrix} 1 & -1 \\ -1 & 1 \end{bmatrix} \begin{matrix} 2 \\ 3 \end{matrix} \text{ lb/in.} \quad (\text{columns } 2, 3)$$

$$[k]^{(3)} = 6.396\times10^6 \begin{bmatrix} 1 & -1 \\ -1 & 1 \end{bmatrix} \begin{matrix} 3 \\ 4 \end{matrix} \text{ lb/in.} \quad (\text{columns } 3, 4)$$

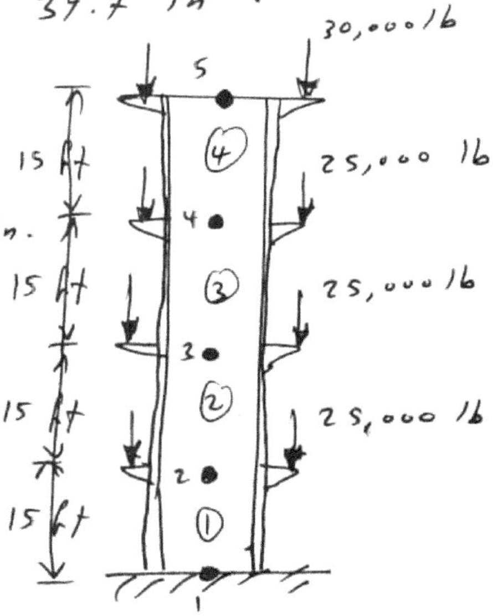

$$[k]^{(4)} = 6.396\times10^{6}\begin{array}{c} \begin{matrix} 4 & \quad 5 \end{matrix} \\ \begin{bmatrix} 1 & -1 \\ -1 & 1 \end{bmatrix} \end{array}\begin{matrix} 4 \\ 5 \end{matrix} \quad \text{lb/in.}$$

Global Stiffness Matrix:

$$[K] = \begin{matrix} 6.396 \\ \times10^{6} \end{matrix}\begin{array}{c} \begin{matrix} 1 & \quad 2 & \quad 3 & \quad 4 & \quad 5 \end{matrix} \\ \left[\begin{array}{c|c|c|c|c} 1 & -1 & & & \\ \hline -1 & 1+1 & -1 & & \\ \hline & -1 & 1+1 & -1 & \\ \hline & & -1 & 1+1 & -1 \\ \hline & & & -1 & 1 \end{array}\right] \end{array}\begin{matrix} 1 \\ 2 \\ 3 \\ 4 \\ 5 \end{matrix}$$

Boundary conditions: $\boxed{u_1 = 0}$.

$$[K]\{u\} = \{F\}$$

$$6.396\times10^{6}\begin{bmatrix} 1 & -1 & 0 & 0 & 0 \\ -1 & 2 & -1 & 0 & 0 \\ 0 & -1 & 2 & -1 & 0 \\ 0 & 0 & -1 & 2 & -1 \\ 0 & 0 & 0 & -1 & 1 \end{bmatrix}\begin{Bmatrix} 0 \\ u_2 \\ u_3 \\ u_4 \\ u_5 \end{Bmatrix} = \begin{Bmatrix} R_1 \\ 5000 \\ 5000 \\ 5000 \\ 6098 \end{Bmatrix}$$

$\Longrightarrow$ Solve to get:

$$\begin{Bmatrix} u_1 \\ u_2 \\ u_3 \\ u_4 \\ u_5 \end{Bmatrix} = \begin{Bmatrix} 0 \\ 0.03283 \\ 0.05784 \\ 0.07504 \\ 0.08442 \end{Bmatrix} \text{ in.}$$

Axial Stresses:

$$\sigma^{(1)} = \frac{E(u_i - u_j)}{\ell} = -5289 \text{ lb/in}^2.$$

$$\sigma^{(2)} = -4029 \text{ lb/in}^2.$$

$$\sigma^{(3)} = -2771 \text{ lb/in}^2.$$

$$\sigma^{(4)} = -1511 \text{ lb/in}^2.$$

# Chapter 5
# Two-Dimensional Elements

## 5.1 Rectangular Elements:

page 189

The displacement distribution over the element depends on both $X$- and $Y$-coordinates.

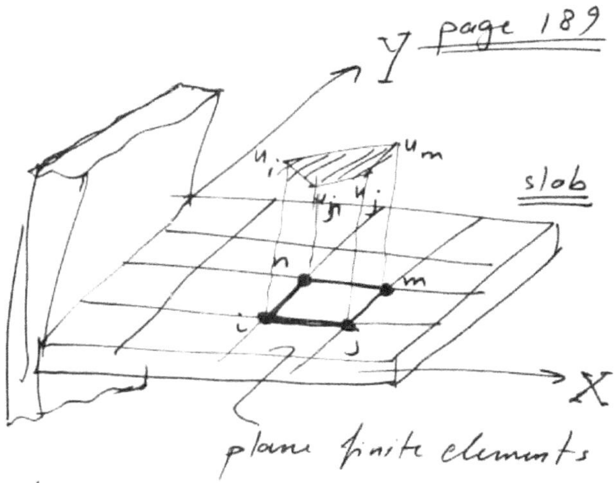


$$u^{(e)} = b_1 + b_2 x + b_3 y + b_4 x y$$

four unknowns because we have four nodes.

The function varies linearly along the edges of the element. Nonlinear only inside the element

$\Rightarrow$ "bilinear element"

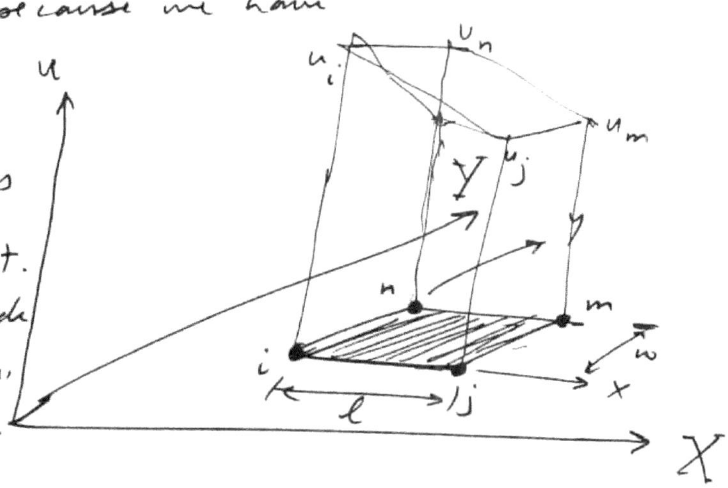


at $x=0, y=0$, $u^{(e)} = u_i$

at $x=l, y=0$, $u^{(e)} = u_j$

at $x=l, y=w$, $u^{(e)} = u_m$

at $x=0, y=w$, $u^{(e)} = u_n$

$x, y$ : local axes.

$X, Y$ : global axes.

$\Rightarrow \quad u_i = b_1 + 0 \Rightarrow u_i = b_1 \Rightarrow b_1 = u_i$

$\quad u_j = b_1 + b_2 l \Rightarrow b_2 = \dfrac{u_j - u_i}{l}$.

$$u_m = b_1 + b_2 l + b_3 w + b_4 l w$$

$$u_n = b_1 + b_3 w \implies b_3 = \frac{u_n - u_i}{w}$$

$$u_m = \cancel{u_i} + (u_j - \cancel{u_i}) + (u_n - u_i) + b_4 l w$$

$$u_m + u_i - u_j - u_n = b_4 l w \implies b_4 = \frac{u_i - u_j + u_m - u_n}{l w}.$$

$$\implies u^{(e)} = S_i u_i + S_j u_j + S_m u_m + S_n u_n$$

$$= [S_i \quad S_j \quad S_m \quad S_n] \begin{Bmatrix} u_i \\ u_j \\ u_m \\ u_n \end{Bmatrix}$$

where $S_i = \left(1 - \frac{x}{l}\right)\left(1 - \frac{y}{w}\right)$

$$S_j = \frac{x}{l}\left(1 - \frac{y}{w}\right)$$

$$S_m = \frac{xy}{lw}$$

$$S_n = \frac{y}{w}\left(1 - \frac{x}{l}\right)$$

---

page 192

**Natural Coordinates:**

↳ one dimensionless local coordinates.

Use $\xi$ for $x$ and $\eta$ for $y$:

$$\boxed{\xi = \frac{2x}{l} - 1}, \qquad \boxed{\eta = \frac{2y}{w} - 1}$$

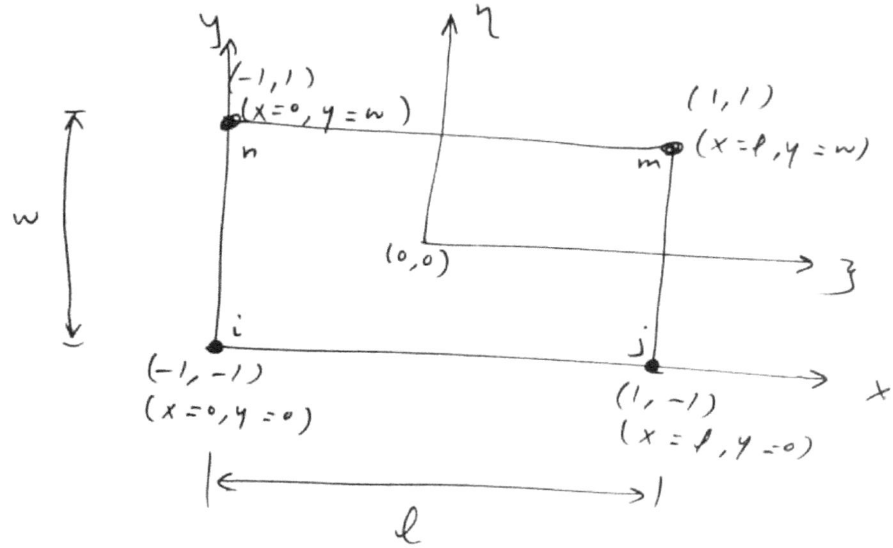


counterclockwise order for numbering.

the bilinear shape functions in natural coordinates:

$$S_i = \frac{1}{4}(1-\xi)(1-\eta)$$
$$S_j = \frac{1}{4}(1+\xi)(1-\eta)$$
$$S_m = \frac{1}{4}(1+\xi)(1+\eta)$$
$$S_n = \frac{1}{4}(1-\xi)(1+\eta)$$

Same properties as before.

$S_i = 1$ at node $i$.

$S_i = 0$ at nodes $j, m, n$.

page 193

## 5.2 Quadratic Quadrilateral Elements:

8-node quadratic quadrilateral element.

* better suited for modeling problems with curved boundaries.

$$u^{(e)} = b_1 + b_2\xi + b_3\eta + b_4\xi\eta + b_5\xi^2 + b_6\eta^2 + b_7\xi^2\eta + b_8\xi\eta^2$$

solve 8 equations in 8 unknowns.

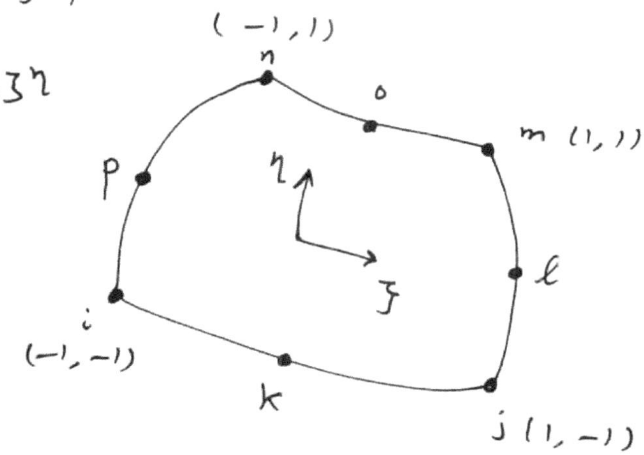

In general, $S = F_1(\xi,\eta)\, F_2(\xi,\eta)$ (S: shape function)

Notes:

① select $F_1$ such that it will produce a value of zero when evaluated along the sides of the element that the given node does not contact.

② select $F_2$ such that when multiplied by $F_1$, it will produce a value of unity at the given node and a value of zero at other neighboring nodes.

Node m: $\xi = 1$, $\eta = 1$.

select $F_1$ such that along $ij$ $(\eta = -1)$ and in $(\xi = -1)$, we will get a value of zero.

$$F_1(\xi,\eta) = (1+\xi)(1+\eta)$$

select $F_2(\xi,\eta) = c_1 + c_2\xi + c_3\eta$

$S_m = F_1 F_2$ should give 1 at m and 0 at $l$ & o.

$S_m = 1$ at $\xi = 1, \eta = 1$ (node m)

$S_m = 0$ at $\xi = 1, \eta = 0$ (node $l$)

$S_m = 0$ at $\xi = 0, \eta = 1$ (node o)

$$S_m = (1+\xi)(1+\eta)(c_1 + c_2\xi + c_3\eta)$$

node m: $1 = (1+1)(1+1)(c_1 + c_2 + c_3)$ ——— (1)

node $l$: $0 = (1+1)(1+0)(c_1 + c_2 + 0)$ ——— (2)

node o: $0 = (1+0)(1+1)(c_1 + 0 + c_3)$ ——— (3)

$$\Longrightarrow \boxed{\begin{aligned} 4c_1 + 4c_2 + 4c_3 &= 1 \\ 2c_1 + 2c_2 &= 0 \\ 2c_1 + 2c_3 &= 0 \end{aligned}}$$

$\Rightarrow \quad c_2 = -c_1$

$c_3 = -c_1$

$\Rightarrow \quad 4c_1 - 4c_1 - 4c_1 = 1 \Rightarrow -4c_1 = 1 \Rightarrow c_1 = -\frac{1}{4}$.

$c_2 = \frac{1}{4}$

$c_3 = \frac{1}{4}$.

$$\therefore S_m = (1+\xi)(1+\eta)\left(-\tfrac{1}{4} + \tfrac{1}{4}\xi + \tfrac{1}{4}\eta\right)$$

Similarly, we get:

$$S_i = -\tfrac{1}{4}(1-\xi)(1-\eta)(1+\xi+\eta)$$

$$S_j = \tfrac{1}{4}(1+\xi)(1-\eta)(-1+\xi-\eta)$$

$$S_m = \tfrac{1}{4}(1+\xi)(1+\eta)(-1+\xi+\eta)$$

$$S_n = -\tfrac{1}{4}(1-\xi)(1+\eta)(1+\xi-\eta)$$

} corner nodes

For the middle nodes:

Node o : $\xi = 0, \ \eta = 1$.

select $F_1$ such that it is zero on sides $ij$ ($\eta = -1$), $in$ ($\xi = -1$), $im$ ($\xi = 1$).

$$\therefore F_1(\xi,\eta) = (1-\xi)(1+\eta)(1+\xi).$$

will produce linear and nonlinear terms.

$\Rightarrow F_2$ must be a constant.

$$F_2(\xi,\eta) = c_1$$

$S_o = 1$ at o ($\xi = 0, \eta = 1$)

$\Rightarrow c_1(1-0)(1+1)(1+0) = 1 \Rightarrow 2c_1 = 1 \Rightarrow c_1 = \frac{1}{2}$.

$$\Rightarrow S_o = \tfrac{1}{2}(1-\xi)(1+\eta)(1+\xi).$$

similarly,

$$S_k = \tfrac{1}{2}(1-\eta)(1-\xi^2).$$

$$S_\ell = \tfrac{1}{2}(1+\xi)(1-\eta^2).$$

$$S_o = \tfrac{1}{2}(1+\eta)(1-\xi)^2$$

$$S_p = \tfrac{1}{2}(1-\xi)(1-\eta^2).$$

---

page 195

**Example 5.1:**

We have used two-dimensional rectangular elements to model the stress distribution in a thin plate. The nodal stresses for an element belonging to the plate are shown in the figure. What is the value of the stress at the center of this element?

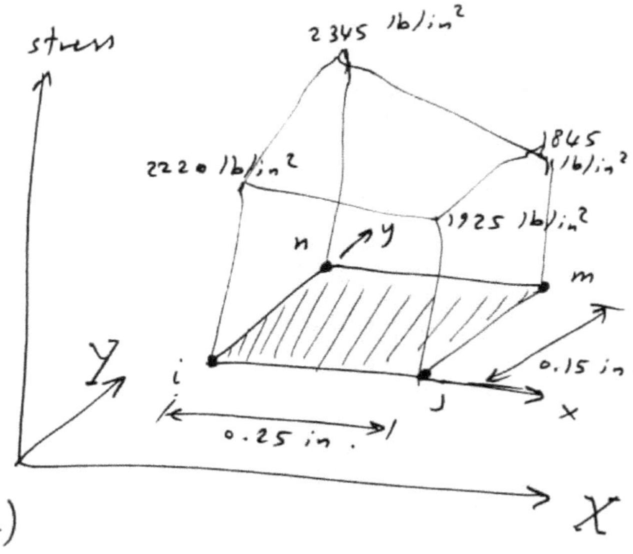


**Solution:**

$$\sigma^{(e)} = [S_i \;\; S_j \;\; S_m \;\; S_n]\begin{Bmatrix}\sigma_i\\ \sigma_j\\ \sigma_m\\ \sigma_n\end{Bmatrix}$$

$$S_i = \left(1-\frac{x}{\ell}\right)\left(1-\frac{y}{w}\right)$$

$$= \left(1-\frac{x}{0.25}\right)\left(1-\frac{y}{0.15}\right)$$

$$S_j = \frac{x}{\ell}\left(1-\frac{y}{w}\right) = \frac{x}{0.25}\left(1-\frac{y}{0.15}\right)$$

$$S_m = \frac{xy}{\ell w} = \frac{xy}{(0.25)(0.15)}$$

$$S_n = \frac{y}{w}\left(1-\frac{x}{\ell}\right) = \frac{y}{0.15}\left(1-\frac{x}{0.25}\right)$$

$$\therefore \sigma^{(e)} = \left(1-\frac{x}{0.25}\right)\left(1-\frac{y}{0.15}\right)(2220) + \frac{x}{0.25}\left(1-\frac{y}{0.15}\right)(1975)$$

$$+ \frac{xy}{(0.25)(0.15)}(1845) + \frac{y}{0.15}\left(1-\frac{x}{0.25}\right)(2345)$$

---

At the <u>midpoint</u> of the element,

$x = 0.125$ in, $y = 0.075$ in.

substitute in the above equation:

$$\therefore \sigma^{(e)}_{midpoint} = 555 + 481 + 461 + 586 = \underline{\underline{2083}} \ \text{lb/in}^2.$$

---

Using natural coordinates is easier: ($\xi = 0$, $\eta = 0$ at midpoint)

$$S_i = \tfrac{1}{4}(1-\xi)(1-\eta) = \tfrac{1}{4}.$$

$$S_j = \tfrac{1}{4}(1+\xi)(1-\eta) = \tfrac{1}{4}.$$

$$S_m = \tfrac{1}{4}(1+\xi)(1+\eta) = \tfrac{1}{4}.$$

$$S_n = \tfrac{1}{4}(1-\xi)(1+\eta) = \tfrac{1}{4}.$$

$$\therefore \sigma^{(e)}_{midpoint} = \tfrac{1}{4}(2220) + \tfrac{1}{4}(1925) + \tfrac{1}{4}(1845) + \tfrac{1}{4}(2345)$$

$$= \underline{\underline{2083}} \ \text{lb/in}^2.$$ (<u>average</u> of nodal stresses).

---

<u>Example 5.2</u>: <u>page 197</u>

Confirm the expression given for the quadratic quadrilateral shape function $S_n$.

<u>Solution</u>: Let $S_n = F_1(\xi,\eta)\, F_2(\xi,\eta)$

select $F_1$ to be zero on sides $ij$ ($\eta = -1$) and $jm$ ($\xi = 1$):

$$\therefore F_1(\xi,\eta) = (1-\xi)(1+\eta).$$

$\therefore$ $F_2$ should be linear.

Let $F_2(\xi,\eta) = c_1 + c_2\xi + c_3\eta$.

at node n, $S_n = 1$ :

$$(1-\xi)(1+\eta)(c_1 + c_2\xi + c_3\eta)\Big|_{\xi=-1,\eta=1} = 1 .$$

$$(2)(2)(c_1 - c_2 + c_3) = 1$$

$$\Rightarrow \quad 4c_1 - 4c_2 + 4c_3 = 1 \quad \text{——} \ (1)$$

at node p $(\xi = -1, \eta = 0)$ , $S_n = 0$ .

$$(2)(1)(c_1 - c_2 + 0) = 0 \quad \Rightarrow \quad 2c_1 - 2c_2 = 0 \quad \text{——} \ (2)$$

at node o $(\xi = 0, \eta = 1)$, $S_n = 0$ .

$$(1)(2)(c_1 + 0 + c_3) = 0 \quad \Rightarrow \quad 2c_1 + 2c_3 = 0 \quad \text{——} \ (3)$$

Solve :

$$\left.\begin{array}{l} c_2 = c_1 \\ c_3 = -c_1 \end{array}\right\} \Rightarrow \quad 4c_1 - 4c_1 - 4c_1 = 1$$

$$-4c_1 = 1 \quad \Rightarrow \quad c_1 = -\tfrac{1}{4} .$$

$$c_2 = -\tfrac{1}{4}, \quad c_3 = \tfrac{1}{4} .$$

$$\therefore \quad S_n = (1-\xi)(1+\eta)\left(-\tfrac{1}{4} - \tfrac{1}{4}\xi + \tfrac{1}{4}\eta\right)$$

$$\therefore \quad S_n = \tfrac{1}{4}(1-\xi)(1+\eta)(-1-\xi+\eta) .$$

$$= -\tfrac{1}{4}(1-\xi)(1+\eta)(1+\xi-\eta) .$$

## 5.3 Linear Triangular Elements:

* A major disadvantage of using bilinear rectangular elements is that they do not conform to a curved boundary very well

* Triangular elements are better suited to approximate curved boundaries.

$$u^{(e)} = a_1 + a_2 X + a_3 Y$$

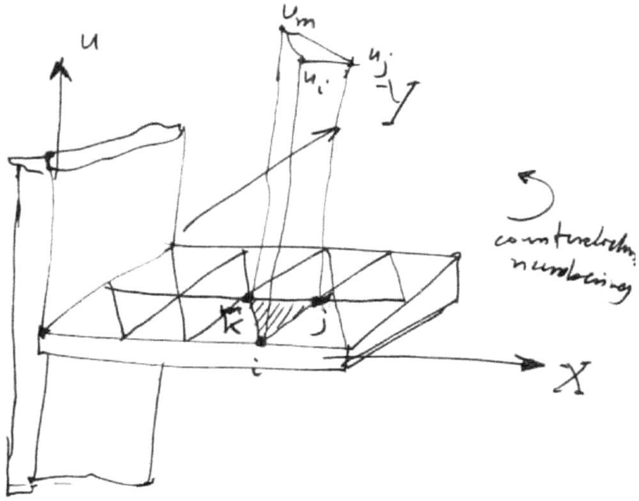


at node i, $u = u_i$,
$X = X_i$, $Y = Y_i$.

at node j, $u = u_j$,
$X = X_j$, $Y = Y_j$.

at node k, $u = u_k$,
$X = X_k$, $Y = Y_k$.

$$\left.\begin{aligned} u_i &= a_1 + a_2 X_i + a_3 Y_i \\ u_j &= a_1 + a_2 X_j + a_3 Y_j \\ u_k &= a_1 + a_2 X_k + a_3 Y_k \end{aligned}\right\} \text{ solve for } a_1, a_2, a_3:$$

$\Longrightarrow$

$$a_1 = \frac{1}{2A}\left[(X_j Y_k - X_k Y_j) u_i + (X_k Y_i - X_i Y_k) u_j + (X_i Y_j - X_j Y_i) u_k\right].$$

$$a_2 = \frac{1}{2A}\left[(Y_j - Y_k) u_i + (Y_k - Y_i) u_j + (Y_i - Y_j) u_k\right].$$

$$a_3 = \frac{1}{2A}\left[(X_k - X_j) u_i + (X_i - X_k) u_j + (X_j - X_i) u_k\right].$$

where $2A = X_i (Y_j - Y_k) + X_j (Y_k - Y_i) + X_k (Y_i - Y_j)$

$A \equiv$ area of triangular element.

$$\therefore \; u^{(e)} = [S_i \;\; S_j \;\; S_k] \begin{Bmatrix} u_i \\ u_j \\ u_k \end{Bmatrix}$$

$$S_i = \frac{1}{2A}(\alpha_i + \beta_i X + \delta_i Y)$$

$$S_j = \frac{1}{2A}(\alpha_j + \beta_j X + \delta_j Y)$$

$$S_k = \frac{1}{2A}(\alpha_k + \beta_k X + \delta_k Y)$$

where

| $\alpha_i = X_j Y_k - X_k Y_j$ | $\beta_i = Y_j - Y_k$ | $\delta_i = X_k - X_j$ |
|---|---|---|
| $\alpha_j = X_k Y_i - X_i Y_k$ | $\beta_j = Y_k - Y_i$ | $\delta_j = X_i - X_k$ |
| $\alpha_k = X_i Y_j - X_j Y_i$ | $\beta_k = Y_i - Y_j$ | $\delta_k = X_j - X_i$ |

$S_i = 1$ at node $i$, $S_i = 0$ at nodes $j$ - $k$.

$\Rightarrow \quad S_i + S_j + S_k = 1$.

---

## Natural (Area) Coordinates for Triangular Elements: page 200

Consider point $P(x, y)$ inside the triangle.

$A_1 + A_2 + A_3 = A$.

Define the area coordinates:
(natural coordinates)

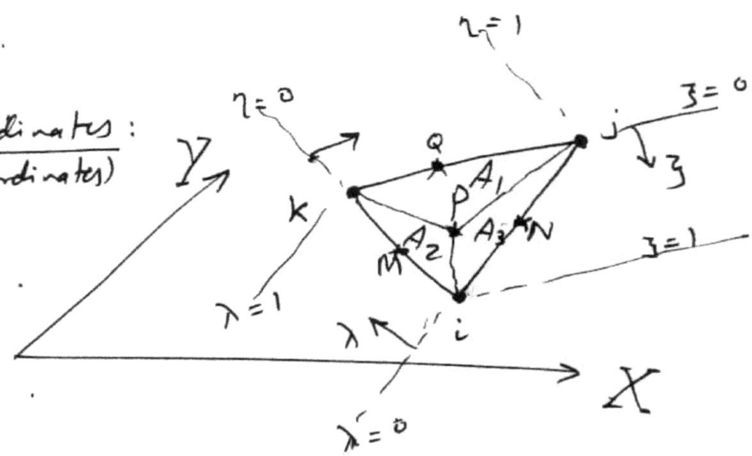


$\xi = \frac{A_1}{A}$, $\xi = 0, \ldots, 1$.

$\eta = \frac{A_2}{A}$, $\eta = 0, \ldots, 1$.

$\lambda = \frac{A_3}{A}$, $\lambda = 0, \ldots, 1$.

168

Only two of the natural coordinates are linearly independent:

$$\frac{A_1}{A} + \frac{A_2}{A} + \frac{A_3}{A} = \frac{A}{A} = \boxed{1 = \xi + \eta + \lambda}$$

e.g. $\lambda = 1 - \xi - \eta$.

Show that the shape functions are:

$$S_i = \xi$$
$$S_j = \eta$$
$$S_k = \lambda$$

e.g. $$\xi = \frac{A_1}{A} = \frac{\frac{1}{2}\left[(X_j Y_k - X_k Y_j) + X(Y_j - Y_k) + Y(X_k - X_j)\right]}{\frac{1}{2}\left[X_i(Y_j - Y_k) + X_j(Y_k - Y_i) + X_k(Y_i - Y_j)\right]}$$

Compare with the previous equations for $S_i$.

---

## 5.4 Quadratic Triangular Elements: page 202

$$u^{(e)} = a_1 + a_2 X + a_3 Y + a_4 X^2 + a_5 XY + a_6 Y^2$$

$$S_i = \xi(2\xi - 1)$$
$$S_j = \eta(2\eta - 1)$$
$$S_k = \lambda(2\lambda - 1)$$
$$= 1 - 3(\xi + \eta) + 2(\xi + \eta)^2.$$
$$S_l = 4\xi\eta.$$
$$S_m = 4\eta\lambda = 4\eta(1 - \xi - \eta).$$
$$S_n = 4\xi\lambda = 4\xi(1 - \xi - \eta).$$

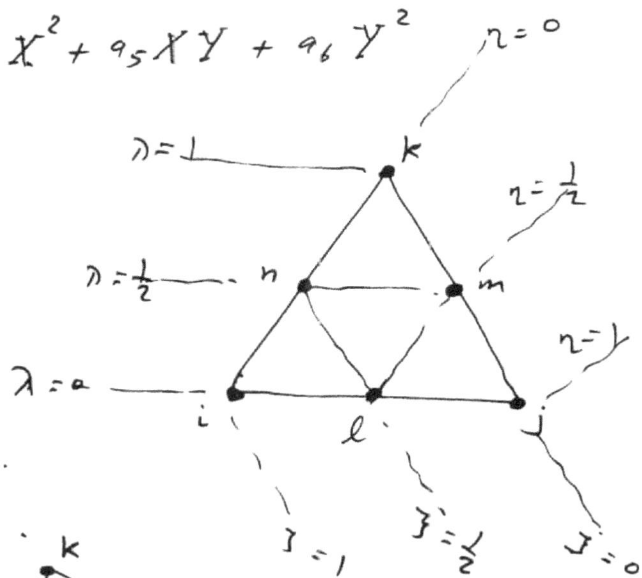


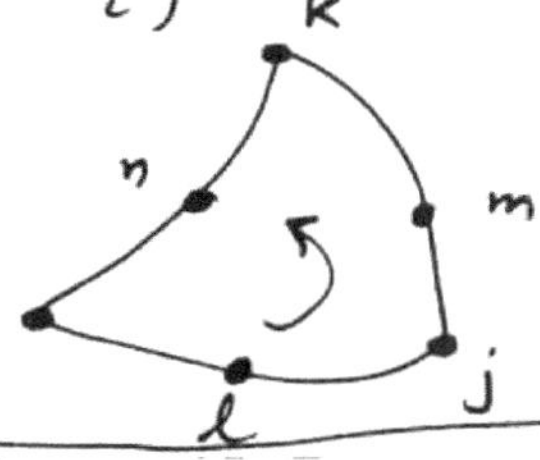


---

Example 5.3 :

We have used two-dimensional triangular elements to model the temperature distribution in a fin. the nodal temperatures and their corresponding positions for an element are shown in the figure.

(a) What is the value of the temperature at $X = 2.15$ cm and $Y = 1.1$ cm ?

(b) Determine the components of temperature gradients for this element ?

(c) Determine the location of $70°C$ and $75°C$ isotherms.

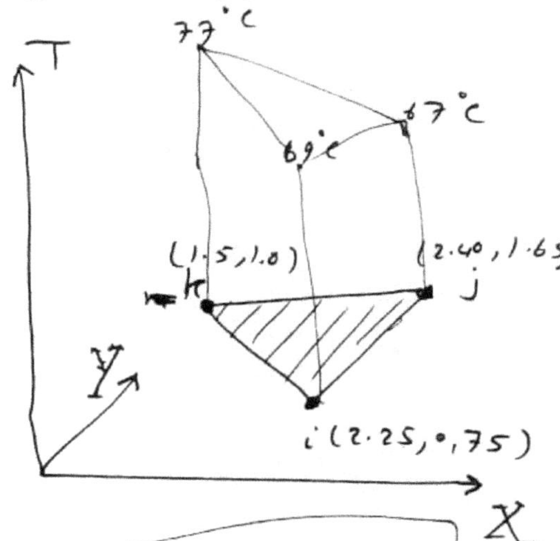


Solution :

(a)

$$T^{(e)} = [S_i \quad S_j \quad S_k] \begin{Bmatrix} T_i \\ T_j \\ T_k \end{Bmatrix}$$

$$= S_i T_i + S_j T_j + S_k T_k .$$

$T_i = 69°C$.
$T_j = 67°C$.
$T_k = 77°C$.

$$S_i = \frac{1}{2A}(\alpha_i + \beta_i X + \delta_i Y).$$

$$S_j = \frac{1}{2A}(\alpha_j + \beta_j X + \delta_j Y)$$

$$S_k = \frac{1}{2A}(\alpha_k + \beta_k X + \delta_k Y)$$

where

$$\alpha_i = X_j Y_k - X_k Y_j = (2.4)(1.0) - (1.5)(1.65) = -0.075 .$$

$$\alpha_j = X_k Y_i - X_i Y_k = (1.5)(0.75) - (2.25)(1.0) = -1.125 .$$

$$\alpha_k = X_i Y_j - X_j Y_i = (2.25)(1.65) - (2.40)(0.75) = 1.9125 .$$

$$\beta_i = Y_j - Y_k = 1.65 - 1.0 = 0.65 .$$

$$\beta_j = Y_k - Y_i = 1.0 - 0.75 = 0.25$$

$$\beta_k = Y_i - Y_j = 0.75 - 1.65 = -0.90$$

$$\delta_i = X_k - X_j = 1.5 - 2.40 = -0.90 .$$

$$\delta_j = X_i - X_k = 2.25 - 1.5 = 0.75$$

$$\delta_k = X_j - X_i = 2.40 - 2.25 = 0.15$$

$$2A = X_i(Y_j - Y_k) + X_j(Y_k - Y_i) + X_k(Y_i - Y_j)$$

$$= 2.25(1.65 - 1.0) + 2.40(1.0 - 0.75) + 1.5(0.75 - 1.65)$$

$$= 0.7125 .$$

$$\therefore S_i = \frac{1}{2A}(\alpha_i + \beta_i X + \delta_i Y) = \frac{1}{0.7125}(-0.075 + 0.65X - 0.90Y) .$$

$$S_j = \frac{1}{2A}(\alpha_j + \beta_j X + \delta_j Y) = \frac{1}{0.7125}(-1.125 + 0.25X + 0.75Y) .$$

$$S_k = \frac{1}{2A}(\alpha_k + \beta_k X + \delta_k Y) = \frac{1}{0.7125}(1.9125 - 0.90X + 0.15Y) .$$

$$\therefore T^{(e)} = S_i T_i + S_j T_j + S_k T_k$$

$$= \frac{69}{0.7125}(-0.075 + 0.65X - 0.90Y) + \frac{67}{0.7125}(-1.125 + 0.25X + 0.75Y)$$

$$+ \frac{77}{0.7125}(1.9125 - 0.90X + 0.15Y) .$$

$$T^{e}_{(X,Y)} = 93.632 - 10.808X - 0.421Y .$$

for the point $X = 2.15$, $Y = 1.1$

$$\Longrightarrow T^{(e)} = 69.93\,^{\circ}C .$$

---

(b) Gradient components:

$$\frac{\partial T^{(e)}}{\partial X} = \frac{\partial}{\partial X}(S_i T_i + S_j T_j + S_k T_k)$$

$$= \frac{1}{2A}(\beta_i T_i + \beta_j T_j + \beta_k T_k)$$

$$\frac{\partial T^{(e)}}{\partial Y} = \frac{1}{2A}(\delta_i T_i + \delta_j T_j + \delta_k T_k)$$

} strains $\varepsilon_x, \varepsilon_y$ for solid mechanics problems.

$$\begin{Bmatrix} \frac{\partial T^{(e)}}{\partial X} \\ \frac{\partial T^{(e)}}{\partial Y} \end{Bmatrix} = \frac{1}{2A} \begin{bmatrix} \beta_i & \beta_j & \beta_k \\ \delta_i & \delta_j & \delta_k \end{bmatrix} \begin{Bmatrix} T_i \\ T_j \\ T_k \end{Bmatrix}$$ = constant values. (independent of X, Y).

⇒ "constant strain triangle".

$$= \frac{1}{0.7125} \begin{bmatrix} 0.65 & 0.25 & -0.9 \\ -0.9 & 0.75 & 0.15 \end{bmatrix} \begin{Bmatrix} 69 \\ 67 \\ 77 \end{Bmatrix} = \begin{Bmatrix} -10.808 \\ -0.421 \end{Bmatrix}$$

at any location inside the element.

---

(c) Temperature varies linearly in both X- and Y- directions.

70°C isotherm (70°C constant temperature line):

$$\frac{77-70}{77-69} = \frac{1.5-X}{1.5-2.25} = \frac{1.0-Y}{1.0-0.75}$$

⇒ $X = 2.16$ cm, $Y = 0.78$ cm.

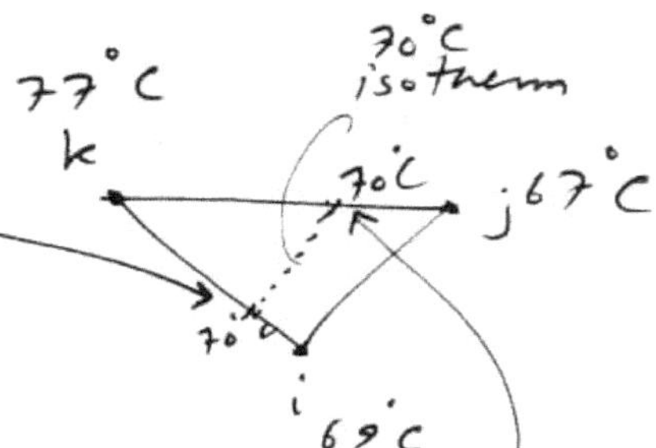


$$\frac{77-70}{77-67} = \frac{1.5-X}{1.5-2.4} = \frac{1.0-Y}{1.0-1.65}$$

⇒ $X = 2.13$ cm, $Y = 1.45$ cm.

---

75°C isotherm:

$$\frac{77-75}{77-69} = \frac{1.5-X}{1.5-2.25} = \frac{1.0-Y}{1.0-0.75}$$

⇒ $X = 1.69$ cm, $Y = 0.94$ cm

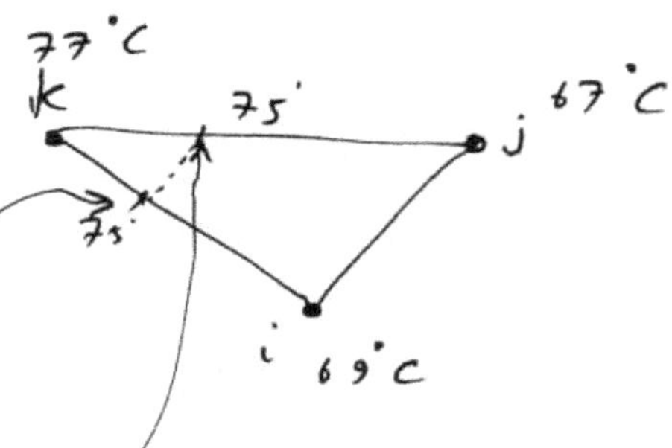


$$\frac{77-75}{77-67} = \frac{1.5-X}{1.5-2.4} = \frac{1.0-Y}{1.0-1.65}$$

⇒ $X = 1.68$ cm, $Y = 1.13$ cm.

## 5.5 Isoparametric Elements:

An isoparametric formulation: when we use the same set of shape functions to define the unknown variables $u, v, T, \ldots,$ and the geometry $x, X$.

$$u = S_i U_{ix} + S_j U_{jx} + S_m U_{mx} + S_n U_{nx}$$

$$v = S_i U_{iy} + S_j U_{jy} + S_m U_{my} + S_n U_{ny}$$

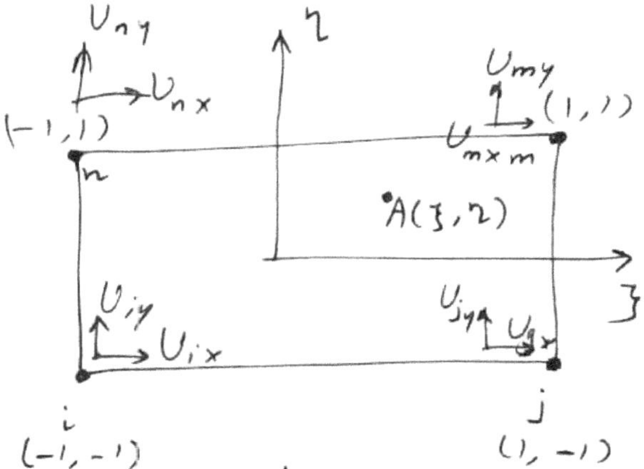


$$\Rightarrow \begin{Bmatrix} u \\ v \end{Bmatrix}_{2\times 1} = \begin{bmatrix} S_i & 0 & S_j & 0 & S_m & 0 & S_n & 0 \\ 0 & S_i & 0 & S_j & 0 & S_m & 0 & S_n \end{bmatrix}_{2\times 8} \begin{Bmatrix} U_{ix} \\ U_{iy} \\ U_{jx} \\ U_{jy} \\ U_{mx} \\ U_{my} \\ U_{nx} \\ U_{ny} \end{Bmatrix}_{8\times 1}$$

Using the isoparametric formulation, we can write:

$$x = S_i x_i + S_j x_j + S_m x_m + S_n x_n$$

$$y = S_i y_i + S_j y_j + S_m y_m + S_n y_n$$

$S_i, S_j, S_m, S_n$ are written in terms of $\xi, \eta$.

$$\frac{\partial f(x,y)}{\partial \xi} = \frac{\partial f(x,y)}{\partial x}\frac{\partial x}{\partial \xi} + \frac{\partial f(x,y)}{\partial y}\frac{\partial y}{\partial \xi}$$

$$\frac{\partial f(x,y)}{\partial \eta} = \frac{\partial f(x,y)}{\partial x}\frac{\partial x}{\partial \eta} + \frac{\partial f(x,y)}{\partial y}\frac{\partial y}{\partial \eta}$$

We need this change of coordinates for the derivatives.

$$\begin{Bmatrix} \frac{\partial f(x,y)}{\partial \xi} \\ \frac{\partial f(x,y)}{\partial \eta} \end{Bmatrix} = \begin{bmatrix} \frac{\partial x}{\partial \xi} & \frac{\partial y}{\partial \xi} \\ \frac{\partial x}{\partial \eta} & \frac{\partial y}{\partial \eta} \end{bmatrix} \begin{Bmatrix} \frac{\partial f(x,y)}{\partial x} \\ \frac{\partial f(x,y)}{\partial y} \end{Bmatrix}$$

let $[J] = \begin{bmatrix} \frac{\partial x}{\partial \xi} & \frac{\partial y}{\partial \xi} \\ \frac{\partial x}{\partial \eta} & \frac{\partial y}{\partial \eta} \end{bmatrix}$ ≡ Jacobian of the coordinate transformation.

$$\Rightarrow \begin{Bmatrix} \frac{\partial f(x,y)}{\partial x} \\ \frac{\partial f(x,y)}{\partial y} \end{Bmatrix} = [J]^{-1} \begin{Bmatrix} \frac{\partial f(x,y)}{\partial \xi} \\ \frac{\partial f(x,y)}{\partial \eta} \end{Bmatrix}$$

To be continued in Chapter 8.

---

## 5.6 Two-Dimensional Integrals: Gauss-Legendre Quadrature:

page 207

$$I = \int_{-1}^{+1}\int_{-1}^{+1} f(\xi,\eta)\,d\xi\,d\eta \cong \int_{-1}^{+1}\left[\sum_{i=1}^{n} w_i f(\xi_i,\eta)\right] d\eta$$

$$= \sum_{i=1}^{n}\sum_{j=1}^{n} w_i w_j f(\xi_i,\eta_j)$$

---

**Example 5.4:**

page 208

Evaluate the integral $I = \int_0^2\int_0^2 (3y^2+2x)\,dx\,dy$.

**Solution:**

Exact solution:

$$I = \int_0^2\int_0^2 (3y^2+2x)\,dx\,dy = \int_0^2\left[\int_0^2 (3y^2+2x)\,dx\right] dy$$

$$= \int_0^2 \left[3y^2x + x^2\right]_0^2 dy = \int_0^2 (6y^2+4)\,dy = \left[6\frac{y^3}{3} + 4y\right]_0^2$$

$$= \frac{48}{3} + 4(2) = \boxed{24}.$$

Using Gauss-Legendre quadrature:

change variables from $x, y \rightarrow \xi, \eta$:

$$x = \left(\frac{b+a}{2}\right) + \left(\frac{b-a}{2}\right)\lambda \qquad \text{from chapter 3.}$$

$$x = \left(\frac{2+0}{2}\right) + \left(\frac{2-0}{2}\right)\xi = 1+\xi \implies \begin{matrix} x = 1+\xi \\ dx = d\xi \end{matrix}$$

$$y = \left(\frac{2+0}{2}\right) + \left(\frac{2-0}{2}\right)\eta = 1+\eta \implies \begin{matrix} y = 1+\eta \\ dy = d\eta \end{matrix}$$

$$\therefore I = \int_0^2\int_0^2 (3y^2 + 2x)\,dx\,dy = \int_{-1}^{1}\int_{-1}^{1}\left[3(1+\eta)^2 + 2(1+\xi)\right]d\xi\,d\eta$$

$$\cong \sum_{i=1}^{n}\sum_{j=1}^{n} w_i w_j f(\xi_i, \eta_j) \qquad \xi_i, \eta_i = \text{sampling points.}$$

$$\cong \sum_{i=1}^{2}\sum_{j=1}^{2} w_i w_j \left[3(1+\eta_j)^2 + 2(1+\xi_i)\right] \qquad \text{using the two-point formula.}$$

= sum first over $i$ (twice) then over $j$ (twice)

$w_i = w_j = 1$

$\xi_i = \eta_i = \pm 0.577350269$

$= 24.00000000$.

---

## 5.7 Examples of Two-Dimensional Elements in ANSYS:

Linear & Quadratic quadrilateral & triangular elements. page 209

① **Plane 2**: six-node triangular structural-solid element.

quadratic shape functions.

$x, y$ at each node.

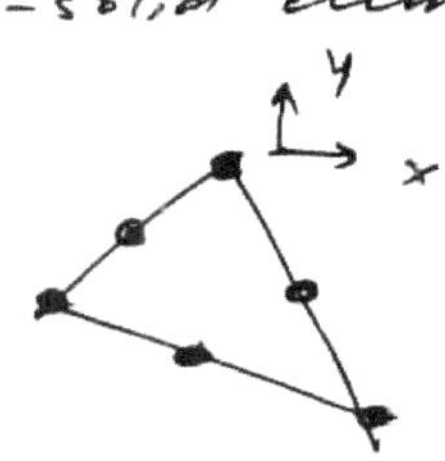

② Plane 42 : four-node quadrilateral element for structural-solid problems.

x, y at each node.

③ Plane 82 : eight-node quadrilateral element for structural-solid problems.

x, y at each node.

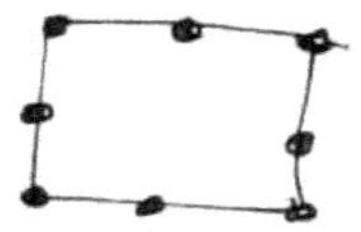

* more suited for curved boundaries.
* requires more computational time.

---

Homework 5 pages 210 – 215 :

problems 6, 7, 8, 9, 11, 18, 24.

---

www.ingramcontent.com/pod-product-compliance
Lightning Source LLC
LaVergne TN
LVHW082358200926
848481LV00012B/3603